Communications
in Computer and Information Science   1858

Rationale

The CCIS series is devoted to the publication of proceedings of computer science conferences. Its aim is to efficiently disseminate original research results in informatics in printed and electronic form. While the focus is on publication of peer-reviewed full papers presenting mature work, inclusion of reviewed short papers reporting on work in progress is welcome, too. Besides globally relevant meetings with internationally representative program committees guaranteeing a strict peer-reviewing and paper selection process, conferences run by societies or of high regional or national relevance are also considered for publication.

Topics

The topical scope of CCIS spans the entire spectrum of informatics ranging from foundational topics in the theory of computing to information and communications science and technology and a broad variety of interdisciplinary application fields.

Information for Volume Editors and Authors

Publication in CCIS is free of charge. No royalties are paid, however, we offer registered conference participants temporary free access to the online version of the conference proceedings on SpringerLink (http://link.springer.com) by means of an http referrer from the conference website and/or a number of complimentary printed copies, as specified in the official acceptance email of the event.

CCIS proceedings can be published in time for distribution at conferences or as post-proceedings, and delivered in the form of printed books and/or electronically as USBs and/or e-content licenses for accessing proceedings at SpringerLink. Furthermore, CCIS proceedings are included in the CCIS electronic book series hosted in the SpringerLink digital library at http://link.springer.com/bookseries/7899. Conferences publishing in CCIS are allowed to use Online Conference Service (OCS) for managing the whole proceedings lifecycle (from submission and reviewing to preparing for publication) free of charge.

Publication process

The language of publication is exclusively English. Authors publishing in CCIS have to sign the Springer CCIS copyright transfer form, however, they are free to use their material published in CCIS for substantially changed, more elaborate subsequent publications elsewhere. For the preparation of the camera-ready papers/files, authors have to strictly adhere to the Springer CCIS Authors' Instructions and are strongly encouraged to use the CCIS LaTeX style files or templates.

Abstracting/Indexing

CCIS is abstracted/indexed in DBLP, Google Scholar, EI-Compendex, Mathematical Reviews, SCImago, Scopus. CCIS volumes are also submitted for the inclusion in ISI Proceedings.

How to start

To start the evaluation of your proposal for inclusion in the CCIS series, please send an e-mail to ccis@springer.com.

Ana Fred · Carlo Sansone · Oleg Gusikhin ·
Kurosh Madani

Editors

Deep Learning Theory and Applications

Third International Conference, DeLTA 2022
Lisbon, Portugal, July 12–14, 2022
Revised Selected Papers

Springer

Editors
Ana Fred
Instituto de Telecomunicações and University
of Lisbon
Lisbon, Portugal

Oleg Gusikhin
Ford Motor Company
Commerce Township, MI, USA

Carlo Sansone
University of Napoli Federico II
Napoli, Italy

Kurosh Madani
Univ Paris-Est Creteil (UPEC)
Créteil, France

ISSN 1865-0929   ISSN 1865-0937 (electronic)
Communications in Computer and Information Science
ISBN 978-3-031-37316-9   ISBN 978-3-031-37317-6 (eBook)
https://doi.org/10.1007/978-3-031-37317-6

This Springer imprint is published by the registered company Springer Nature Switzerland AG
The registered company address is: Gewerbestrasse 11, 6330 Cham, Switzerland

Preface

The present book includes extended and revised versions of a set of selected papers from the 3rd International Conference on Deep Learning Theory and Applications (DeLTA 2022), held in Lisbon, Portugal, from 12–14 July 2022.

DeLTA 2022 received 36 paper submissions from 21 countries, of which 17% were included in this book.

The papers were selected by the event chairs and their selection is based on a number of criteria that include the classifications and comments provided by the program committee members, the session chairs' assessment and also the program chairs' global view of all papers included in the technical program. The authors of selected papers were then invited to submit a revised and extended version of their papers having at least 30% innovative material.

Deep Learning and Big Data Analytics are two major topics of data science, nowadays. Big Data has become important in practice, as many organizations have been collecting massive amounts of data that can contain useful information for business analysis and decisions, impacting existing and future technology. A key benefit of Deep Learning is the ability to process these data and extract high-level complex abstractions as data representations, making it a valuable tool for Big Data Analytics, where raw data is largely unlabeled.

Machine learning and artificial intelligence are pervasive in most real-world application scenarios such as computer vision, information retrieval and summarization from structured and unstructured multimodal data sources, natural language understanding and translation, and many other application domains. Deep learning approaches, leveraging on big data, are outperforming state-of-the-art more "classical" supervised and unsupervised approaches, directly learning relevant features and data representations without requiring explicit domain knowledge or human feature engineering. These approaches are currently highly important in IoT applications.

The papers selected and included in this book contribute to the understanding of relevant trends of current research on Deep Learning Theory and Applications, focusing on: Image Classification, Convolutional Neural Networks (CNN), Clustering, Classification and Regression, Unsupervised Feature Learning, Semantic Segmentation, IoT and Smart Devices, Active Learning, Graph Representation Learning, Dimensionality Reduction and Deep Reinforcement Learning.

We would like to thank all the authors for their contributions and also the reviewers who have helped to ensure the quality of this publication.

July 2022

Ana Fred
Carlo Sansone
Oleg Gusikhin
Kurosh Madani

Organization

Conference Co-chairs

Oleg Gusikhin   Ford Motor Company, USA
Kurosh Madani   University of Paris-Est Créteil, France

Program Co-chairs

Ana Fred   Instituto de Telecomunicações and University of
     Lisbon, Portugal
Carlo Sansone   University of Naples Federico II, Italy

Program Committee

Enrico Blanzieri   University of Trento, Italy
Marco Buzzelli   University of Milano - Bicocca, Italy
Claudio Cusano   University of Pavia, Italy
Shyam Diwakar   Amrita University, India
Ke-Lin Du   Concordia University, Canada
Gilles Guillot   CSL Behring/Swiss Institute for Translational and
     Entrepreneurial Medicine, Switzerland
Chih-Chin Lai   National University of Kaohsiung,
     Taiwan, Republic of China
Chang-Hsing Lee   Chung Hua University, Taiwan, Republic of China
Marco Leo   National Research Council of Italy, Italy
Fuhai Li   Washington University Saint Louis, USA
Huaqing Li   Southwest University, China
Xingyu Li   University of Alberta, Canada
Yung-Hui Li   National Central University,
     Taiwan, Republic of China
Perry Moerland   Amsterdam UMC, University of Amsterdam, The
     Netherlands
Tomoyuki Naito   Osaka University, Japan
Le-Minh Nguyen   Japan Advanced Institute of Science and
     Technology, Japan
Juan Pantrigo   Universidad Rey Juan Carlos, Spain

Oksana Pomorova	University of Lodz, Poland
Mircea-Bogdan Radac	Politehnica University of Timisoara, Romania
Sivaramakrishnan Rajaraman	National Library of Medicine, USA
Jitae Shin	Sungkyunkwan University, South Korea
Sunghwan Sohn	Mayo Clinic, USA
Minghe Sun	University of Texas at San Antonio, USA
Ryszard Tadeusiewicz	AGH University of Science and Technology, Poland
Jayaraman Valadi	Shiv Nadar University, India
Aalt van Dijk	Wageningen University & Research, The Netherlands
Theodore Willke	Intel Corporation, USA
Jianhua Xuan	Virginia Tech, USA
Seokwon Yeom	Daegu University, South Korea
Yizhou Yu	University of Hong Kong, China

Additional Reviewers

Ganlong Zhao	University of Hong Kong

Invited Speakers

Ioannis Pitas	Aristotle University of Thessaloniki, Greece
Michal Irani	Weizmann Institute of Science, Israel
João Freitas	PagerDuty, Portugal

Contents

Modified SkipGram Negative Sampling Model for Faster Convergence of Graph Embedding

Kostas Loumponias[✉][iD], Andreas Kosmatopoulos[iD], Theodora Tsikrika[iD], Stefanos Vrochidis[iD], and Ioannis Kompatsiaris[iD]

Information Technologies Institute, Centre for Research and Technology Hellas-CERTH, 54124 Thessaloniki, Greece
{loumponias,akosmato,theodora.tsikrika,stefanos,ikom}@iti.gr

Abstract. Graph embedding techniques have been introduced in recent years with the aim of mapping graph data into low-dimensional vector spaces, so that conventional machine learning methods can be exploited. In particular, in the DeepWalk model, truncated random walks are employed in random walk-based approaches to capture structural links-connections between nodes. The SkipGram model is then applied to the truncated random walks to compute the embedded nodes. In this work, the proposed DeepWalk model provides a faster convergence speed than the standard one by introducing a new trainable parameter in the model. Furthermore, experimental results on real-world datasets show that the performance in downstream community detection and link prediction task is improved by using the proposed DeepWalk model.

Keywords: Graph embedding · DeepWalk · Community detection · Link prediction

1 Introduction

In the past few years, there has been a significant increase in the volume of data generated by services that utilise various type of networks. Graphs analysis is used for representing information in various networks (e.g., citation networks, sensor networks, social networks [8] etc.) as graphs, taking into account the interactions between the network entities. Consequently, inherent properties of the network (a.k.a graph) can be discovered, using graph analytical tasks, such as node classification [2], community detection [15], link prediction [18] and visualization [20].

Recently, graph embedding methods that represent graph nodes in a vector space have been developed. The main goal of graph embedding methods is to map graph nodes into a low-dimensional latent vector space, while maximally preserving the properties of the graph structure. Therefore, node similarity in the original complex irregular spaces can be quantified based on various similarity measures in the latent vector space (or embedded space). In addition, more accurate graph analytics tasks can be leveraged from the learned embedded space, as opposed to directly performing such tasks in the

A. Fred et al. (Eds.): DeLTA 2022, CCIS 1858, pp. 1–16, 2023.
https://doi.org/10.1007/978-3-031-37317-6_1

high-dimensional complex graph domain. Graph embedding methods can be classified into three main categories [3,8]: (i) factorization-based, (ii) random walk-based, and (iii) deep learning-based.

Factorization-based methods describe the connections between nodes as a matrix and factorize this matrix to obtain the embedded nodes. The most common matrices used to represent the connections between nodes are the node adjacency matrix, Laplacian matrix, and node transition probability matrix. Based on the characteristics of the representative matrix, different approaches to factorization might be used. In the case of the Laplacian matrix, eigenvalue decomposition can be used if the obtained matrix is positive semi-definite. Gradient descent algorithms can be used to speed up the factorization-based methods.

Random walk-based methods are used to obtain the topological relationships between nodes by performing truncated random walks. To that end, a graph is converted into a collection of node sequences (using truncated random walks), in which the frequency of node pairs measures the structural distance between them. Then, machine-learning (ML) based methods are used for obtaining the embedding. The most common method used to calculate the embedded nodes using truncated random walks is the *SkipGram model* [21].

Deep learning-based methods apply well-established deep learning (DP) models on a whole graph (or the corresponding proximity matrix) to obtain the embedded nodes. Autoencoders have been utilised for dimensionality reduction [23] due to their ability to model non-linear structure in the data. Furthermore, as an extension of the standard Convolutional Neural Network, the Graph Convolution Neural Network [29] has been proposed to deal with non-Euclidian structural data, such as the graphs.

In this work, we focus on random walk-based methods, proposing a novel SkipGram model that provides a faster convergence speed in calculating the embedded nodes. The proposed approach introduces a new function, called *sigmoid b* $\sigma_b(x)$, in order to tackle some of the limitations of the typical sigmoid function, such as the saturated values [11]. Subsequently, the forward and back-propagation stage of the proposed SkipGram model are calculated and the calculated embedded nodes are utilised in the community and link prediction tasks. More precisely, the k-means algorithm [10] is applied to embedded nodes to calculate the network communities, while the logistic regression model [12] is used to predict the existence of edges (links) between two nodes in the graph.

This work extends our previous work [18] as follows: First of all, proofs of the proposed SkipGram model (not included in [18]) are provided in detail, along with a more comprehensive description of the standard SkipGram model. Moreover, further extensive evaluation experiments are performed for demonstrating the effectiveness of the proposed method by considering additional real-world datasets and by examining in depth the effect of the different hyperparameters of the proposed model on the performance for the community detection and node classification tasks.

The rest of the paper is organised as follows: In Sect. 2, random walk-based methods are described. In Sect. 3, the proposed method is provided. In Sect. 4, experimental results are presented using real-world networks to demonstrate the effectiveness of the proposed framework. Finally, in Sect. 5, conclusions and future work are discussed.

2 Related Work

The DeepWalk (DW) method [25] adopts the SkipGram model, which is a neural language model for producing graph embeddings. More specifically, the SkipGram model attempts to maximize the likelihood of words that appear within the same sentence. Thus, DW first performs truncated random walks (with fixed length t) for each node of the graph to obtain a set of node sequences. This process is repeated n times (number of walks) and it follows that a node and a node sequence can be interpreted as a word and a sentence, respectively. Then, the SkipGram model is applied on the node sequence to maximize the likelihood of observing a node's neighbourhood conditioned on its embedded nodes. Hence, nodes with similar neighbourhoods (second order proximity) share similar embeddings.

In the same way as in DW, the node2vec (n2v) method [9] preserves higher-order proximity between nodes using truncated random walks (with fixed length) and the SkipGram model. The main difference between DW and node2vec is that n2v employs biased-random walks that provide a trade-off between breadth-first (BFS) and depth-first (DFS) graph searches. The results have shown that in many network tasks, such as community detection and node classification tasks, n2v produces higher-quality and more informative nodes embedded than DW.

In the DW and n2v methods, the embedded nodes are randomly initialized. However, such initializations may end up trapped in local optima, since the objective function of DW and n2v is non-convex. In order to tackle this limitation, hierarchical representation learning for networks (HARP) [5] was proposed, where it provides a novel method to initialize the model weights. HARP aims to find a smaller graph which approximates the global structure of the original graph. This simplified graph is utilised to learn a set of initial representations, which serve as good initializations for learning representations in the original graph. HARP is a general meta-strategy to improve all graph embedding methods, such as DW and n2v.

Moreover, the WALKLETS method [26] generates multi-scale representations of graph nodes, by sub-sampling short random walks on the nodes. By skipping some nodes, WALKLETS alters the random walk method used in DW and n2v. To that end, a similar to factorizing GraRep approach [4] is performed for multiple skip lengths. The resulting node sequences are used for training the SkipGram model. Finally, other variations of the above methods are Deep Random Walk [17] and Tri-party Deep Network Representation [24].

3 Graph Embedding: Random Walk Based Technique

In this section, a brief description of the DW method is provided, considering the negative sampling approach [22] in SkipGram model. Next, the proposed SkipGram model is presented, providing detailed proofs. Following that, the proposed SkipGram model is used as an initial step for the community detection and link prediction downstream tasks.

3.1   Standard SkipGram Model: Negative Sampling Approach

The DW (and n2v) method consists of two steps, (i) random walk sampling and (ii) the SkipGram model (see Fig. 1). In this work, we focus on the SkipGram model, thus, no additional details about random walks are reported from here on. Let $G = (V, E)$ be a graph, where V and $E \subseteq (V \times V)$ stands for the node and edge set of graph G, respectively. The standard SkipGram model corresponds to a fully connected neural network with one hidden layer (without any activation function) and multiple outputs (see Fig. 1). The main goal of SkipGram model is to predict the surrounding nodes (context nodes) of a given target node. To that end, the embedded vectors of the nodes are calculated.

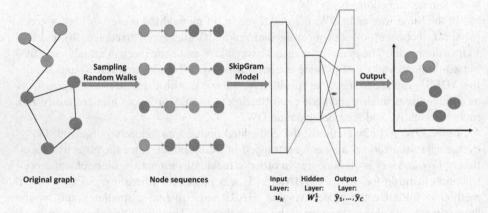

Fig. 1. The DeepWalk model.

Next, the notation of SkipGram model parameters are provided and described in detail. The input vector $\mathbf{u}_k$ is the one-hot vector of target node $u_k \in V$, $\mathbf{W}^1 \in M_{|V| \times d}$[1] is the embedded matrix, where $|V|$ denotes the total number of the nodes and d the embedding size. Each row $(i = 1, 2, ..., |V|)$ of $\mathbf{W}^1$ represents the embedded vector of node $u_i \in V$. $\mathbf{W}^2 \in M_{d \times |V|}$ is the output embedded matrix, while $\{\hat{\mathbf{y}}_{k-w}, ..., \hat{\mathbf{y}}_{k-1}, \hat{\mathbf{y}}_{k+1}, ..., \hat{\mathbf{y}}_{k+w}\}$ are the predicted context nodes (one-hot vectors) when the input-target node is u_k, (or equivalently $\mathbf{u}_k$), where w denotes the window size. The sets of predictions $\{\hat{\mathbf{y}}_{k+i}\}_{i=-w:w, i \neq 0}$ for convenience will be denoted as $\{\hat{\mathbf{y}}_1, ..., \hat{\mathbf{y}}_C\}$.

From now on, it assumed that $\mathbf{W}^1_s = \mathbf{W}^{1'} \cdot \mathbf{u}_k$, where $\mathbf{W}^1_s$ corresponds to the s-row of $\mathbf{W}^1$, since $\mathbf{u}_k$ is one-hot-vector. Furthermore, $\mathbf{W}^1_s$ stands for the embedded vector of target node u_k. In the standard SkipGram model, the cost function for the embedded target $\mathbf{W}^1_s$ is calculated as

$$J(\theta) = -\sum_{c=1}^{C} \log \frac{\exp\left(\mathbf{W}^{2'}_c \cdot \mathbf{W}^1_s\right)}{\sum_{i=1}^{|V|} \exp\left(\mathbf{W}^{2'}_i \cdot \mathbf{W}^1_s\right)}, \tag{1}$$

---

[1] $M_{n \times m}$ denotes the set of matrices $n \times m$.

where $\theta = [\mathbf{W}^1, \mathbf{W}^2]$ and $\mathbf{W}_c^2$ represents the c-th column of $\mathbf{W}^2$. In addition, the function

$$P(u_c|u_k; \theta) = \frac{\exp\left(\mathbf{W}_c^{2'} \cdot \mathbf{W}_s^1\right)}{\sum_{i=1}^{|V|} \exp\left(\mathbf{W}_i^{2'} \cdot \mathbf{W}_s^1\right)} \tag{2}$$

represents the conditional probability of observing a context node u_c (embedded output $\mathbf{W}_c^2$) given the target node u_k (embedded target $\mathbf{W}_s^1$). It is clear that the cost function (1) is computationally inefficient, since the denominator in (1) requires $|V|$ iterations (total number of nodes). Due to this computational burden, cost function (1) is not used in most implementations of SkipGram model.

In order to overcome this limitation, the negative sampling process is used, which reduces the complexity of the SkipGram model. In a nut shell, the negative sampling process draws K number of negative samples (pair of nodes with low proximity) using the noise distribution $P_n(w)$ [22], for each positive pair (pair of nodes with high proximity). Thus, the logarithm of condition probability function (2) is approximated by

$$\log P(u_c|u_k; \theta) = \log \sigma\left(\mathbf{W}_c^{2'} \cdot \mathbf{W}_s^1\right) + \sum_{i=1}^{K} \log \sigma\left(-\mathbf{W}_{neg(i)}^{2'} \cdot \mathbf{W}_s^1\right), \tag{3}$$

where $\sigma(x)$ is the sigmoid function, while the row $\mathbf{W}_{neg(i)}^2$ is randomly selected from matrix $\mathbf{W}^2$, using the noise distribution $P_n(w)$. The first term of (3) indicates the logarithmic probability of u_c (embedded vector $\mathbf{W}_c^2$) to appear within the context window of the target node u_k (embedded vector $\mathbf{W}_s^1$), while, the second term indicates the logarithmic probability of node $u_{neg(i)}$ (embedded vector $\mathbf{W}_{neg(i)}^2$) not appearing in the context window of u_k.

In the negative sampling process, $K+1$ columns of the output embedded matrix $\mathbf{W}^2$ are updated, while in the embedded matrix $\mathbf{W}^1$ only the row $\mathbf{W}_s^1$ is updated, since the input $\mathbf{u}_k$ is one-hot vector. The number of negative samples K usually is set equal to 5. Finally, the update equations of SkipGram model parameters are calculated as follows:

$$\mathbf{c}_j = \mathbf{c}_i - \eta \cdot (\sigma(x_i) - t_i) \cdot \mathbf{W}_s^1, \tag{4}$$

$$\mathbf{W}_s^1 = \mathbf{W}_s^1 - \eta \cdot \sum_{i=1}^{K+1} (\sigma(x_i) - t_i) \cdot \mathbf{c}_i, \tag{5}$$

where $\mathbf{c}_i = \begin{cases} \mathbf{W}_c^2, & i = 1 \\ \mathbf{W}_{neg(j-1)}^2, & i = 2, ..., K+1 \end{cases}$, $t_i = \begin{cases} 1, & i = 1 \\ 0, & i = 2, ..., K+1 \end{cases}$,

$x_i = \mathbf{c}_i' \cdot \mathbf{W}_s^1$ and η is the learning rate.

The term $(\sigma(x_i) - t_i)$ in the update Eqs. (4), (5) is derived from the derivatives of $-\log \sigma(x_i)$ with respect to $\mathbf{c}_i$ and $\mathbf{W}_s^1$ for $i = 1, ..., K+1$, respectively. In the case of positive sample (i.e., $i = 1$) and low values of x_i (i.e., $x_i \rightarrow -\infty$), the term $(\sigma(x_i) - t_i)$ is maximized. Therefore, the SkipGram model updates-corrects the weights $\theta = [\mathbf{W}^1, \mathbf{W}^2]$ for low values of x_i, otherwise, when the values of x_i are high, the updates (values of $(\sigma(x_i) - t_i)$) are negligible. In the case of negative samples ($i \neq 1$) and low values of x_i, the updates are negligible, otherwise for high values of x_i the updates are maximized. Thus, the inner product $x_i = \mathbf{c}_i' \cdot \mathbf{W}_s^1$ defines a proximity between the nodes u_k and u_c, and the aim of the SkipGram model is to maximize and minimize it for positive and negative samples, respectively.

3.2 Proposed SkipGram Model

In the standard negative sampling approach described earlier, the conditional probability function (2) is approximated using the sigmoid function (3). As it is known, the values of a probability function must lie in the interval $[0, 1]$; this makes the sigmoid function an appropriate choice. However, one of the main drawbacks of the sigmoid function is the saturated values of its derivatives (gradients). More specifically, the derivatives of $\log \sigma(x)$ converge to 0 for $x \to +\infty$ and to 1 for $x \to -\infty$. Therefore, for any low value of x, the derivative is essentially equivalent to 1. Thus, the range of the updates is constrained, and its maximum value is 1. This can lead to a significant computing cost until the weights θ converge, according to the gradient descent method.

In order to tackle the above restriction, the sigmoid b function is proposed

$$\sigma_b(x) = \frac{1}{1 + \exp(-b \cdot x)}, \tag{6}$$

where $b > 0$. It is clear that the values of the proposed function lie in the interval $[0, 1]$, since it approximates the conditional probability (3). Next, in Lemma 1 the derivatives of the proposed $\sigma_b(x)$ with respect to (w.r.t.) x and b are provided.

Lemma 1. *The derivatives of* $\log \sigma_b(x)$ *w.r.t.* x *and* b *are equal to:*

$$\frac{\partial \log \sigma_b(x)}{\partial x} = b \cdot (1 - \sigma_b(x)), \forall x \in \mathbb{R}, \tag{7}$$

$$\frac{\partial \log \sigma_b(x)}{\partial b} = x \cdot (1 - \sigma_b(x)). \tag{8}$$

Proof.

$$\frac{\partial \log \sigma_b(x)}{\partial x} = \frac{1}{\sigma_b(x)} \cdot \frac{\partial \sigma_b(x)}{\partial x} = \frac{1}{\sigma_b(x)} \cdot \frac{\partial}{\partial x} \left(\frac{1}{1 + exp(-b \cdot x)} \right)$$

$$= \frac{1}{\sigma_b(x)} \cdot \frac{b \cdot exp(-b \cdot x)}{(1 + exp(-b \cdot x))^2}$$

$$= b \cdot \frac{1}{\sigma_b(x)} \cdot \frac{1}{1 + exp(-b \cdot x)} \cdot \frac{exp(-b \cdot x)}{1 + exp(-b \cdot x)}$$

$$= b \cdot (1 - \sigma_b(x)).$$

since $\sigma_b(x) \neq 0, \forall x \in \mathbb{R}$. Next, in the same way is proved that

$$\frac{\partial \log \sigma_b(x)}{\partial b} = x \cdot (1 - \sigma_b(x)).$$

$\square$

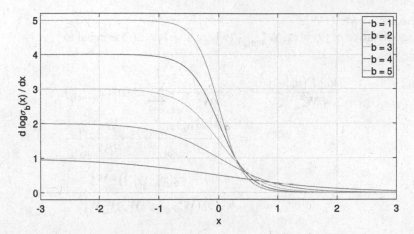

Fig. 2. The derivatives of $\log \sigma_b(x)$.

Figure 2 illustrates the derivative of $\log \sigma_b(x)$ (Eq. 7) w.r.t. x for different values of parameter b. It is clear that, in the case of $b = 1$, the derivative of $\log \sigma_b(x)$ is similar to the derivative of $\log \sigma(x)$. Furthermore, the range of updates (derivatives) for $b > 1$ are higher than the corresponding for $b = 1$. Next, in Proposition 1 the update equations of $\mathbf{W}_c^2$, $\mathbf{W}_{neg(i)}^2$, $\mathbf{W}_s^1$ and b using $\sigma_b(x)$ and the gradient descent technique are provided.

Proposition 1. *The update equations of the negative logarithm conditional probability function*

$$- \log P(u_c|u_k; \theta) = - \log \sigma_b \left(\mathbf{W}_c^{2'} \cdot \mathbf{W}_s^1 \right) - \sum_{i=1}^{K} \log \sigma_b \left(-\mathbf{W}_{neg(i)}^{2'} \cdot \mathbf{W}_s^1 \right), \quad (9)$$

w.r.t. $\mathbf{W}_c^2$, $\mathbf{W}_{neg(i)}^2$, $\mathbf{W}_s^1$ *and* b *using the gradient descent technique are*

$$\boldsymbol{c}_i = \boldsymbol{c}_i - \eta \cdot b \cdot (\sigma_b(x_i) - t_i) \cdot \mathbf{W}_s^1, \quad (10)$$

$$\mathbf{W}_s^1 = \mathbf{W}_s^1 - \eta \cdot b \cdot \sum_{i=1}^{K+1} (\sigma_b(x_i) - t_i) \cdot \boldsymbol{c}_i, \quad (11)$$

and

$$b = b - \eta_b \cdot \sum_{i=1}^{K+1} x_i \cdot (\sigma_b(x_i) - t_i), \quad (12)$$

where η_b *is the learning rate of parameter* b.

Proof. Initially, the derivatives of (9) w.r.t. $\mathbf{W}_c^2$, $\mathbf{W}_{neg(i)}^2$, $\mathbf{W}_s^1$ and b are calculated.

The derivative of (9) w.r.t. $\mathbf{W}_c^2$ is equal to

$$- \frac{\partial \log P(u_c|u_k; \theta)}{\partial \mathbf{W}_c^2} = - \frac{\partial \log \sigma_b(x_c)}{\partial \mathbf{W}_c^2} = - \frac{\partial \log \sigma_b(x_c)}{\partial x_c} \cdot \frac{\partial x_c}{\partial \mathbf{W}_c^2}$$

$$\overset{(7)}{=} -b \cdot (1 - \sigma_b(x_c)) \cdot \mathbf{W}_s^1 = b \cdot (\sigma_b(x_c) - 1) \cdot \mathbf{W}_s^1 \quad (13)$$

where $x_c = \mathbf{W}_c^{2'} \cdot \mathbf{W}_s^1$.

The derivative of (9) w.r.t. $\mathbf{W}_{neg(j)}^2$ (where $j = 1, ..., K$) is equal to

$$
\begin{aligned}
-\frac{\partial \log P(u_c|u_k; \theta)}{\partial \mathbf{W}_{neg(j)}^2} &= -\frac{\partial}{\partial \mathbf{W}_{neg(j)}^2} \sum_{i=1}^K \log \sigma_b(x_{neg(i)}^-) \\
&= -\frac{\partial \log \sigma_b(x_{neg(j)}^-)}{\partial x_{neg(j)}^-} \cdot \frac{\partial x_{neg(j)}^-}{\partial \mathbf{W}_{neg(j)}^2} \\
&\stackrel{(7)}{=} b \cdot (1 - \sigma_b(x_{neg(j)}^-)) \cdot \mathbf{W}_s^1 \\
&= b \cdot (\sigma_b(\mathbf{W}_{neg(j)}^{2'} \cdot \mathbf{W}_s^1)) \cdot \mathbf{W}_s^1
\end{aligned}
\tag{14}
$$

where $x_{neg(j)}^- = -\mathbf{W}_{neg(j)}^{2'} \cdot \mathbf{W}_s^1$.

The derivative of (9) w.r.t. $\mathbf{W}_s^1$ is equal to

$$
\begin{aligned}
-\frac{\partial \log P(u_c|u_k; \theta)}{\partial \mathbf{W}_s^1} &= -\frac{\partial \log \sigma_b(x_c)}{\partial \mathbf{W}_s^1} - \frac{\partial}{\partial \mathbf{W}_s^1} \sum_{i=1}^K \log \sigma_b(x_{neg(i)}^-) \\
&= \frac{\partial \log \sigma_b(x_c)}{\partial x_c} \cdot \frac{\partial x_c}{\partial \mathbf{W}_s^1} - \sum_{i=1}^K \frac{\partial \log \sigma_b(x_{neg(i)}^-)}{\partial x_{neg(i)}^-} \cdot \frac{\partial x_{neg(i)}^-}{\partial \mathbf{W}_s^1} \\
&\stackrel{(7)}{=} -b \cdot (1 - \sigma_b(x_c)) \cdot \mathbf{W}_c^2 + \sum_{i=1}^K b \cdot (1 - \sigma_b(x_{neg(i)}^-)) \cdot \mathbf{W}_{neg(i)}^2 \\
&= \sum_{i=1}^{K+1} b \cdot (\sigma_b(x_i) - t_i) \cdot \mathbf{c}_i,
\end{aligned}
\tag{15}
$$

where the definitions of x_i, t_i and $\mathbf{c}_i$ are provided and described in subsection 3.1.

Next, the objective function (9) can be written as

$$
-\log P(u_c|u_k; \theta) = -\log \sigma_b(x_1) - \sum_{i=1}^K \log \sigma_b(-x_{i+1}),
$$

therefore, the derivative of the above objective function w.r.t. b is calculated as:

$$-\frac{\partial \log P(u_c|u_k; \theta)}{\partial b} = -\frac{\partial \log \sigma_b(x_1)}{\partial b} - \sum_{i=1}^{K} \frac{\partial \log \sigma_b(-x_{i+1})}{\partial b}$$

$$\overset{(8)}{=} -x_1 \cdot (1 - \sigma_b(x_1)) + \sum_{i=1}^{K} x_{i+1} \left(1 - \sigma_b(-x_{x+i})\right)$$

$$= x_1 \cdot (\sigma_b(x_1) - 1) + \sum_{i=1}^{K} x_{i+1} \left(\sigma_b(x_{i+i}) - 0\right)$$

$$= \sum_{i=1}^{K+1} x_i \left(\sigma_b(x_i) - t_i\right). \tag{16}$$

Finally, using the gradient descent technique [13] and the derivatives (13)–(16), the update equations of the proposed SkipGram model are calculated.   □

The results in Proposition 1 show that the proposed model (SkipGram$_b$) and $\sigma_b(x)$ can be used to replace the standard one (SkipGram and $\sigma(x)$) in any machine learning model. However, the updates Eqs. (10)–(12) must be adjusted to the particular machine learning model.

It is important to note that numerous activation functions have been presented in literature in an effort to improve performance in various scientific domains. The SM-Taylor softmax function was used in [1] for image classification tasks and the results showed that it performed better than the standard softmax function. In addition, some of the best known and well established alternative activation functions are the Exponential Linear Unit (ELU) [6] and Scaled exponential Linear Unit (SELU) [14]. However, the aforementioned functions cannot be applied to embedding methods, since the function must be bounded in the interval [0, 1] as previously explained.

3.3 Community Detection and Link Prediction Tasks

The proposed DeepWalk$_b$ (DW$_b$) method has similar framework to standard DW, with the main difference being the use of the SkipGram$_b$ model instead of the standard one. In Algorithm 1, the proposed framework for community detection is presented. More precisely, lines 1–7 represent the DW$_b$ process. Then, the $k-$means algorithm is applied on the embedded nodes to detect the communities of the graph (*Com*). Moreover, the proposed SkipGram$_b$ model can be also applied to the n2v process, since DW and n2v differ only in how truncated random walks are performed, as shown in line 4 of Algorithm 1.

The process of link prediction and evaluation are provided in Algorithm 2. Three sub-graphs, denoted G_{tr}, G_{mod} and G_{ts} are explicitly derived from the original graph G. To that end, the StellarGraph tool [9] is used to split the graph $G = (V, E)$ to $G_{tr} = (V, E_{tr})$, $G_{mod} = (V, E_{mod})$ and $G_{ts} = (V, E_{ts})$. The embedded nodes θ_{tr} are first calculated using the *train* graph G_{tr}. Then, using four distinct operators (*oprtr*): Hadamard product, $L1$, $L2$ norm [19] and the simple average, the similarities between

---

Algorithm 1. DeepWalk$_b$ for community detection.

Require: Graph G, number of communities k, window size w, embedding size d, walk length t, number of walks n
1: **for** i=1:n **do**
2:     $O = Shuffle(V)$
3:     **for** $u_i \in O$ **do**
4:         $RW_{u_i} = $ RandomWalk(G, u_i, t)
5:         $\theta = $ SkipGram$_b(RW_{u_i}, w)$
6:     **end for**
7: **end for**
8: $Com = k$-means(θ, k)
9: **return** Com

Algorithm 2. DeepWalk$_b$ for link prediction.

Require: Graph G, window size w, embedding size d, walk length t, number of walks n
1: $G_{tr}, G_{mod}, G_{ts} = $ split(G)
2: $\theta_{tr} = $ DW$_b(G_{tr}, w, d, t, n)$
3: **for** i=1:ν_{oprtr} **do**
4:     clsfr(i) = Logistic Regression$(\theta_{tr}$, operator(i)$)$
5:     AccScore(i) = evaluate(clsfr(i), G_{mod}, θ_{tr}, oprtr(i))
6: **end for**
7: $i_{max} = $ argmax(AccScore)
8: $clsfr_{max}, oprtr_{max} = $ clsfr(i_{max}), oprtr(i_{max})
9: $\theta_{ts} = $ DW$_b(G_{tsr}, w, d, t, n)$
10: Test Score = evaluate$(clsfr_{max}, G_{ts}, \theta_{ts}, oprtr_{max})$

---

embedded nodes are calculated. Afterwards, the classifiers, one for each operator, are calculated using the logistic regression model (whether the nodes are connected or not). Then, the classifiers are evaluated considering the *model selection* graph G_{mod} and the embedded nodes θ_{tr}. The operator with the highest accuracy score is used to evaluate the classifier for the *test* graph G_{ts}.

4   Experimental Evaluation

In this section, the proposed method, DW$_b$ is evaluated against the baseline of DW on community detection and link prediction tasks. Implementation details are discussed before presenting results on real-world datasets.

4.1   Data Description and Implementation Details

The proposed method DW$_b$ and DW are evaluated on five publicly available real-world datasets with ground-truth communities: Cora, CiteSeer, PubMed [27], ego-Facebook and Amazon [16]. Cora (2708 nodes and 5429 edges), CiteSeer (3312 nodes and 4732 edges) and PubMed (19717 nodes and 44338 edges) contain publications, ego-Facebook (2871 nodes and 62334 edges) contains social circles formed from users of

Facebook, while Amazon (15,716 nodes and 48739 edges) contains products found in the Amazon website that are frequently bought together. Table 1 showcases the real-world datasets used in the evaluation, as well as the number of their communities.

Table 1. Real-world datasets.

Dataset	Nodes	Edges	Communities
Cora	2,708	5,429	7
CiteSeer	3,312	4,732	6
PubMed	19,717	44,338	3
ego-Facebook	2,871	63,334	147
Amazon	15,716	48,739	1,229

Next, the parameters used in methods DW_b and DW are $w = 10$ (window size), $d = 128$ (embedding size), $t = 80$ (walk-length), while the number of epochs and the batch-size are equal to 15 and 1000, respectively. The learning rate η is set equal to 0.02 for both methods. In [18] it was shown that a larger η value increases the convergence speed of DW with a trade-off in community detection performance. However, DW_b outperforms DW in convergence speed even for larger η values [18]. Thus, the scope of this work is to evaluate DW_b on the learning rate of parameter b. To that end, the experimental sets in DW_b are conducted considering different values of learning rate, $\eta_b \in \{0.01, 0.05, 0.2\}$. Finally, both methods use the stochastic gradient descent technique with momentum 0.9, for the back propagation process.

In each experiment, the Adjusted Rand Index (ARI), the Normalized Mutual Information (NMI) and the graph's modularity (Mod) [28] are calculated for the community detection task, while the Area Under the Curve (AUC) score [7] is calculated for the link prediction task. In addition, the edge sets of the sub-graphs generated by the StellarGraph tool are defined in all experimental sets as follows: E_{ts} includes 90% of the total edges (E), while the E_{tr} and E_{mod} include 75% and 25% of E_{ts}, respectively.

4.2   Evaluation of DeepWalk$_b$ Model

Table 2 details the performances of DW_b (for various values of η_b) and standard DW in community detection and link prediction tasks, where *cd epochs* and *lp epochs* denote the required number of epochs in order the community detection metrics (i.e. ARI, NMI, Mod) and link prediction metric (AUC) to converge. Moreover, Fig. 3 illustrates the performance of DW and DW_b considering the metrics ARI, NMI, Mod and AUC (y axis) for different values of η_b in Cora, CiteSeer, PubMed, eqo-Facebook and Amazon dataset, where the x axis (in the sub-figures) stands for the number of epochs.

As it is expected, DW_b converges faster than the standard DW in all datasets for both tasks due to the trainable parameter b. It is clear that the learning rate η_b has a low impact on convergence speed for DW_b. Only in the Cora and ego-Facebook datasets, (significantly) fewer epochs are required using higher values of η_b in link prediction and community detection task, respectively. Furthermore, the parameter b gets larger values during the training stage, when large values of η_b are used, while when community detection and link predictions metrics converge, the values of b either remain unchanged or decrease (see Fig. 4). This property of parameter b is expected, since when the DW_b model converges, large updates to the model parameters are no longer required.

In addition, the proposed DW_b performs the community detection and link prediction tasks as well as (or better than) standard DW in fewer epochs. The best performances (for both methods) for all metrics within 15 epochs are detailed in Table 2. As it can be seen, DW_b provides the same performances (differences less than 0.02)

Table 2. The best performances of DW and DW_b regarding the metrics ARI, NMI, Mod and AUC. The columns *cd epochs* and *lp epochs* contain the required number of epochs in order for the community detection metrics (i.e. ARI, NMI, Mod) and link prediction metric (AUC), respectively, to converge.

Dataset	Method	ARI	NMI	Mod	AUC	cd epochs	lp epochs
Cora	DW_b $\eta_b = 0.01$	0.389	0.457	0.745	0.893	2	6
	DW_b $\eta_b = 0.05$	0.392	0.463	0.743	0.893	1	2
	DW_b $\eta_b = 0.2$	0.389	0.457	0.741	0.903	2	2
	DW	0.390	0.463	0.747	0.879	7	14
CiteSeer	DW_b $\eta_b = 0.01$	0.127	0.457	0.725	0.913	2	4
	DW_b $\eta_b = 0.05$	0.151	0.462	0.727	0.905	2	3
	DW_b $\eta_b = 0.2$	0.124	0.462	0.737	0.911	1	4
	DW	0.127	0.463	0.701	0.870	8	12
PubMed	DW_b $\eta_b = 0.01$	0.318	0.299	0.602	0.761	1	1
	DW_b $\eta_b = 0.05$	0.319	0.301	0.601	0.771	1	1
	DW_b $\eta_b = 0.2$	0.318	0.299	0.603	0.877	1	1
	DW	0.302	0.296	0.601	0.768	6	6
ego-Facebook	DW_b $\eta_b = 0.01$	0.361	0.647	0.478	0.943	4	6
	DW_b $\eta_b = 0.05$	0.364	0.648	0.461	0.921	3	6
	DW_b $\eta_b = 0.2$	0.361	0.646	0.453	0.923	2	7
	DW	0.318	0.621	0.369	0.892	13	8
Amazon	DW_b $\eta_b = 0.01$	0.569	0.904	0.941	0.995	2	2
	DW_b $\eta_b = 0.05$	0.572	0.904	0.946	0.996	2	1
	DW_b $\eta_b = 0.2$	0.570	0.904	0.943	0.996	1	1
	DW	0.576	0.903	0.820	0.970	11	15

regardless the learning rate η_b for all experimental sets with the exception of two cases. In the CiteSeer dataset, DW_b with $\eta_b = 0.05$ has ARI score equal to 0.151, while the other methods have ARI score close to 0.12. Moreover, in the ego-Facebook dataset, DW_b with $\eta_b = 0.01$ has AUC score equal to 0.943, while the rest DW_b ($\eta_b = 0.05$ and $\eta_b = 0.2$) and DW have AUC score close to 0.92 and 0.89, respectively. Next, the standard DW model provides a poor performance in link prediction task compared to DW_b in all dataset (differences greater than 0.02). Finally, in ego-Facebook dataset, DW_b (regardless of η_b) outperforms DW in all metrics, while providing better Mod. scores on the CiteSeer and Amazon datasets.

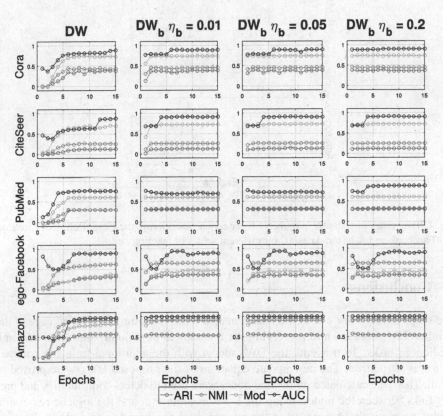

Fig. 3. The performances of DW and DW_b in community detection and link prediction task, considering the Cora, PubMed, CiteSeer, ego-Facebook and Amazon dataset, for different values of learning rate η_b.

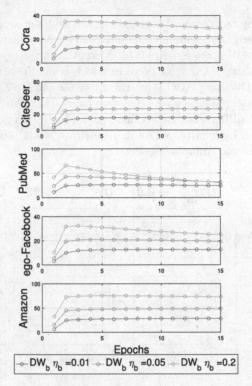

Fig. 4. The values of trainable parameter b per epoch, considering the Cora, PubMed, CiteSeer, ego-Facebook and Amazon datasets, for different values of learning rate η_b.

5   Conclusions

The scope of this work was to propose a method for accelerating the convergence of the standard DW model, while preserving the accuracy in community detection and link predictions tasks. To this end, the DW_b model with the additional trainable parameter b was introduced. The new update equations of the proposed DW_b were proved in detail. Then, the calculated embedded nodes were used to detect communities and predict links between the nodes, using the k-means algorithm and the logistic regression model, respectively. According to the experimental results using real-world datasets, DW_b converged faster than DW in all experimental settings. Additionally, DW_b provided better AUC score than DW on all datasets, as well as better performance on the other metrics (i.e., ARI, NMI, Mod) in most cases. Finally, the experimental results showed that different values of learning rate η_b have a low impact on the convergence speed of DW_b, due to the ability of the proposed method to increase or decrease the values of b in a proper way.

References

1. Banerjee, K., Gupta, R.R., Vyas, K., Mishra, B.: Exploring alternatives to softmax function. In: Proceedings of the 2nd International Conference on Deep Learning Theory and Applications - DeLTA, pp. 81–86. INSTICC, SciTePress (2021). https://doi.org/10.5220/0010502000810086
2. Bhagat, S., Cormode, G., Muthukrishnan, S.: Node classification in social networks. In: Aggarwal, C. (ed.) Social Network Data Analytics, pp. 115–148. Springer, Boston (2011). https://doi.org/10.1007/978-1-4419-8462-3_5
3. Cai, H., Zheng, V.W., Chang, K.C.C.: A comprehensive survey of graph embedding: problems, techniques, and applications. IEEE Trans. Knowl. Data Eng. **30**(9), 1616–1637 (2018)
4. Cao, S., Lu, W., Xu, Q.: Grarep: learning graph representations with global structural information. In: Proceedings of the 24th ACM International on Conference on Information and Knowledge Management, pp. 891–900 (2015)
5. Chen, H., Perozzi, B., Hu, Y., Skiena, S.: Harp: hierarchical representation learning for networks. In: Proceedings of the AAAI Conference on Artificial Intelligence, vol. 32 (2018)
6. Clevert, D.A., Unterthiner, T., Hochreiter, S.: Fast and accurate deep network learning by exponential linear units (elus). arXiv preprint arXiv:1511.07289 (2015)
7. Fawcett, T.: An introduction to roc analysis. Pattern Recogn. Lett. **27**(8), 861–874 (2006)
8. Goyal, P., Ferrara, E.: Graph embedding techniques, applications, and performance: a survey. Knowl.-Based Syst. **151**, 78–94 (2018)
9. Grover, A., Leskovec, J.: node2vec: scalable feature learning for networks. In: Proceedings of the 22nd ACM SIGKDD International Conference on Knowledge Discovery and Data Mining, pp. 855–864 (2016)
10. Hartigan, J.A., Wong, M.A.: Algorithm as 136: a k-means clustering algorithm. J. Roy. Stat. Soc. Ser. C (applied statistics) **28**(1), 100–108 (1979)
11. Hochreiter, S.: The vanishing gradient problem during learning recurrent neural nets and problem solutions. Int. J. Uncertainty Fuzziness Knowl.-Based Syst. **6**(02), 107–116 (1998)
12. Hosmer, Jr, D.W., Lemeshow, S., Sturdivant, R.X.: Applied Logistic Regression, vol. 398. Wiley, Hoboken (2013)
13. Ketkar, Nikhil: Stochastic gradient descent. In: Deep Learning with Python, pp. 111–130. Apress, Berkeley (2017). https://doi.org/10.1007/978-1-4842-2766-4_8
14. Klambauer, G., Unterthiner, T., Mayr, A., Hochreiter, S.: Self-normalizing neural networks. Adv. Neural information processing systems 30 (2017)
15. Kosmatopoulos, A., Loumponias, K., Chatzakou, D., Tsikrika, T., Vrochidis, S., Kompatsiaris, I.: Random-walk graph embeddings and the influence of edge weighting strategies in community detection tasks. In: Proceedings of the 2021 Workshop on Open Challenges in Online Social Networks. pp. 9–13 (2021)
16. Leskovec, J., Krevl, A.: SNAP Datasets: Stanford large network dataset collection. http://snap.stanford.edu/data (Jun 2014)
17. Li, J., Zhu, J., Zhang, B.: Discriminative deep random walk for network classification. In: Proceedings of the 54th Annual Meeting of the Association for Computational Linguistics (Volume 1: Long Papers). pp. 1004–1013 (2016)
18. Loumponias, K., Kosmatopoulos, A., Tsikrika, T., Vrochidis, S., Kompatsiaris, I.: A faster converging negative sampling for the graph embedding process in community detection and link prediction tasks. In: Proceedings of the 3rd International Conference on Deep Learning Theory and Applications - Volume 1: DeLTA, pp. 86–93. INSTICC, SciTePress (2022). DOI: 10.5220/0011142000003277
19. Luo, X., Chang, X., Ban, X.: Regression and classification using extreme learning machine based on l1-norm and l2-norm. Neurocomputing **174**, 179–186 (2016)

20. Van der Maaten, L., Hinton, G.: Visualizing data using t-sne. Journal of machine learning research 9(11) (2008)

21. Mikolov, T., Chen, K., Corrado, G., Dean, J.: Efficient estimation of word representations in vector space. arXiv preprint arXiv:1301.3781 (2013)

22. Mikolov, T., Sutskever, I., Chen, K., Corrado, G.S., Dean, J.: Distributed representations of words and phrases and their compositionality. Advances in neural information processing systems **26**, 3111–3119 (2013)

23. Pan, S., Hu, R., Long, G., Jiang, J., Yao, L., Zhang, C.: Adversarially regularized graph autoencoder for graph embedding. arXiv preprint arXiv:1802.04407 (2018)

24. Pan, S., Wu, J., Zhu, X., Zhang, C., Wang, Y.: Tri-party deep network representation. Network **11**(9), 12 (2016)

25. Perozzi, B., Al-Rfou, R., Skiena, S.: Deepwalk: Online learning of social representations. In: Proceedings of the 20th ACM SIGKDD international conference on Knowledge discovery and data mining. pp. 701–710 (2014)

26. Perozzi, B., Kulkarni, V., Chen, H., Skiena, S.: Don't walk, skip! online learning of multi-scale network embeddings. In: Proceedings of the 2017 IEEE/ACM International Conference on Advances in Social Networks Analysis and Mining 2017. pp. 258–265 (2017)

27. Sen, P., Namata, G., Bilgic, M., Getoor, L., Galligher, B., Eliassi-Rad, T.: Collective classification in network data. AI magazine **29**(3), 93–106 (2008)

28. Vinh, N.X., Epps, J., Bailey, J.: Information theoretic measures for clusterings comparison: Variants, properties, normalization and correction for chance. The Journal of Machine Learning Research **11**, 2837–2854 (2010)

29. Wu, F., Souza, A., Zhang, T., Fifty, C., Yu, T., Weinberger, K.: Simplifying graph convolutional networks. In: International conference on machine learning. pp. 6861–6871. PMLR (2019)

Active Collection of Well-Being and Health Data in Mobile Devices

João Marques[1,2], Francisco Faria[1,2], Rita Machado[1], Heitor Cardoso[1,2], Alexandre Bernardino[1,2], and Plinio Moreno[1,2(✉)]

[1] Instituto Superior Técnico, Universidade de Lisboa, 1049-001 Lisboa, Portugal
{joaojtmarques,franciscofaria00,
heitor.cardoso}@tecnico.ulisboa.pt
[2] Institute for Systems and Robotics, 1049-001 Lisboa, Portugal
{alex,plinio}@isr.tecnico.ulisboa.pt
http://tecnico.ulisboa.pt, https://isr.tecnico.ulisboa.pt

Abstract. In the context of self-reported health, where the subjective perception of the patients is reported through simple yet effective questionnaires, information gathering is very important to obtain consistent and meaningful data analysis. Smart phones are a good tool to gather self-reported variables, but the interaction with the user should go ahead of scheduled notifications. We develop an intelligent notification system that learns by exploration the most adequate time to perform a questionnaire, while using just the answers of the notification messages from the user. We address the smart notification as a Reinforcement Learning (RL) problem, considering several states representations and reward functions for the Upper Confidence Bound, Tabular Q-learning and Deep Q-learning. We evaluate the algorithms on a simulator, followed by an initial prototype where the approach with better performance in simulation is selected for a small pilot. The simulator acts as a person, accepting or discarding the notifications according to the behavior of a three typical users. From the simulation experiments the UCB algorithm showed the most promising results, so we implemented and deployed the RL algorithm in a smartphone application with several users. On this initial pilot with four users, the UCB algorithm was able to find the adequate hours to send notifications for quiz answering.

Keywords: mHealth · Notifications · Machine Learning · Personalization · Reinforcement Learning · Receptivity

1 Introduction

Mobile health (mHealth) takes advantage of mobile devices as tools for reporting information on self-reported health [18] and management [19]. In the case of Noncommunicable diseases (NCDs), also known as chronic conditions [4], such as cancer, diabetes, stroke, and other chronic respiratory or cardiovascular diseases, the ability for patients to employ self-management is very important to reduce the overloading of the healthcare systems [5].

Although mHealth is very promising for self-management, some key factors still restrict the adoption of mHealth, for instance, the lack of standards and regulations, privacy concerns, or the limited guidance and acceptance from traditional healthcare providers. Impact of such self-management approaches requires widespread user adoption and engagement [16].

Phone notifications are able to increase user engagement, having been proven to significantly improve verified compliance, compared to not employing this technique [3]. However, the excessive use of notifications generate negative reactions of the users such as feeling overwhelmed and ignoring the notifications [11]. For these reasons, mHealth applications must function and communicate without burdening the consumer. To provide a better engagement with the user, this work presents an approach for an intelligent notification system that interacts with the user at the most appropriate occasions. Thus, we want to find the policy that leads to notifying the users at the opportune moments. Receptivity from the user's point of view corresponds to the perception that notifications are sent at opportune times, which is not easy to measure and infer in a systematic manner. Nevertheless, the contextual information of the user such as location, motion and time of the day, contribute to the identification of the ideal moment for interactions [11].

Most of current applications employ a basic interaction model, which assumes the availability of the user at any time for engagement with the device. To compensate for this strong assumption, the applications are developed with functionalities/changes such as: Customization of features of the notifications (e.g. presentation, alert type) [12], content [11], and intelligent approaches to deliver them [10]. Therefore, intellegint handle of the notifications is a very relevent issue. Although many studies have been published on this kind of system, most of the existing prior work can be divided into [7]: (i) Detecting transitions between activities, assuming these represent the most opportune timings in a user's routine, and (ii) inferring receptivity from the user's context. Most of these approaches rely on supervised learning for activity detection and recognition, aiming at segment the user types and preferences, even labeling moments as opportune [11]. In this work we aim to develop a personalized notification agent that learns the most appropriate time for filling a questionnaire that gathers health/well-being information.

The agent developed in this work resorts to RL algorithms, which explore by sending notifications at various times of the day (i.e. episode of the algorithm). Reward is provided from the actions of the users: (i) accepting the notification, (ii) ignoring the notification and (iii) answering the questionnaire. Rewards are associated to the user's actions. The simulated user just performs actions (i) and (ii), assuming that action (i) leads to questionnaire filling. The actual users perform actions (i)–(iii), because the implemented application allows the user to interact with the questionnaire application. We explore several reward values and learning algorithms, aiming at selecting the most promising algorithm form the simulated users and then running a pilot with the selected algorithm with real-life users. The RL algorithms considered in this work include the state-less Upper Confidence Bandit and two versions of Q-learning: Tabular and Deep. On the Q-learning algorithms we explore various state representations, which may consider context as the day of the week and rewards from previous episodes.

2 Models for Smart Notification and Users

The main goal of the smart notification system is that the user fills out the required well-being questionnaire. Once the goal is achieved, no more notifications should be sent. The RL agent considers an answered notification as the terminal state, meaning that the episode (in our context, a day), has ended and that the agent only starts working again when the next day begins. The only other terminal state is at the end of the day (24 h). Tasks such as this are called episodic tasks [14].

The agent's main objective is to decide whether to send a notification or stay silent according to the current state and the expected future reward. In the case of UCB, the state is not considered, and the action that maximizes the value function is selected. In the case of Q learning, the action that maximizes the action-value function for the current state is selected. Thus, accepted notifications denote positive reward, while ignored or dismissed notifications correspond to negative reward. The agent's behavior changes accordingly, always intending to increase the long-run sum of rewards (reinforcement signals). We start describing the learning algorithms, followed by the state representation, reward options, and the model that simulates the user's response. We study multiple combination of algorithms, rewards and state representations. This paper extends the work from [9]: (i) considering a more realistic types of responses from the user, (ii) implementing an android app that runs the UCB algorithm for each user and (iii) running a pilot with 4 users to evaluate UCB-day, which was the most promising algorithm from the previous work.

2.1 Reinforcement Learning Algorithms

The Bellman equation is one of the central elements of various RL algorithms [1]. It essentially decomposes the value function into 2 sections, the immediate reward and the discounted future state-action values yet to be received, allowing for a derivation of a relation between state values (V) and action values (Q), as shown in the following:

$$Q_\pi(s,a) = \sum_{s'} P_a\left(s,s'\right)\left(R_a\left(s,s'\right) + \gamma V_\pi\left(s'\right)\right) \tag{1}$$

$$\text{with} \quad V_\pi(s) = \sum_a \pi(s,a) Q_\pi(s,a), \tag{2}$$

where: a - Action; s - State; V - State-value function, defined as the expected empirical reward for each state; π - Policy, representing a mapping of specific states and their respective action values; Q - Action-value function, defined as the expected empirical reward for each action; P_a - Probability of taking action a from state s to new state s'; R_a - Reward from taking action a in state s and moving to state s'; and γ - Discount factor.

The Bellman equation simplifies the computation of the value function in such a way that rather than summing over multiple timesteps, an optimal solution for complex problems can be found by simply breaking it down into smaller, recursive, easier to solve subproblems. Furthermore, it allows the direct approximation of both an optimal action-value function (3), and an optimal state-value function (4).

$$q_*(s,a) = \mathbb{E}\left[R_{t+1} + \gamma \max_{a'}\left(q_*\left(S_{t+1},a'\right)\right) \mid S_t = s, A_t = a\right]$$

$$\Leftrightarrow q_*(s,a) = \sum_{s',r} p\left(s',r \mid s,a\right)\left[r + \gamma \max_{a'}\left(q_*\left(s',a'\right)\right)\right] \quad (3)$$

$$v_*(s) = \max_{a \in A(s)} q_{\pi_*}(s,a)$$

$$v_*(s) = \max_{a} \sum_{s',r} p\left(s',r \mid s,a\right)\left[r + \gamma v_*\left(s'\right)\right] \quad (4)$$

The work in [9] compared several algorithms that aim to compute $q_*(s,a)$ in (3) for the smart notification agent: (i) Upper Confidence Bound [14], (ii) Tabular Q-learning [17] and Deep Q-learning [13].

Upper Confidence Bound (UCB). UCB is the most straightforward RL algorithm, because the environment is modeled as a single state. Thus, the Q function depends only on the selected action [14]. The action is selected according to:

$$a_t \doteq \operatorname*{argmax}_{a}\left[Q_t(a) + c\sqrt{\frac{\ln(t)}{N_t(a)}}\right], \quad (5)$$

where A and a represents action, t represents the current timestep, Q is the action-value function and c, the confidence level that controls the degree of exploration. Finally, $N_t(a)$ corresponds to the number of times action a has been selected before time t. In (5), the term in the square root represents the uncertainty of the estimates of action values, yielding A_t as an upper bound of the probable value of each action. Given a large enough time, UCB executes all the available actions, guaranteeing that the agent explores the action space properly. As time goes on and different actions are performed, to each, the sum of received rewards and the number of selections are associated. With these values, the action-value function Q is updated at each timestep.

To overcome the simplistic state representation, in the previous work we considered multiple instances of UCB that run at different time slots [9]. We assume that time slots model two types of agenda: (i) Hours of day independent of any other information and (ii) Hours of day with its corresponding day of the week. In the first case, the agent will learn to act at each hour of the day (24 instances), while in the second case the agent learn to act at the joined day of the week and hour of the day ($7 \times 24 = 168$ instances).

Tabular Q-learning (TQ). Tabular methods are settled on the core idea of RL in its most straightforward format, that both the state and action spaces are small enough for the estimates of action values and their mapping with specific contexts to be represented as arrays or tables. These methods are commonly used due to their simplicity and ease of computation. Q-learning was initially defined by [17] as follows:

$$Q(S_t, A_t) \leftarrow Q(S_t, A_t) + \alpha[R_{t+1} + \gamma \max_{a}(Q(S_{t+1},a)) - Q(S_t, A_t)], \quad (6)$$

where A and a represent actions, S represents the state and R the reward. Additionally, t represents the current timestep, Q is the action-value function for each state-action pair, α is the learning rate and, lastly, γ is the discount factor. The learning rate, α, determines when Q-values are updated, overriding older information. The discount factor, γ weights the importance of future rewards relative to the more immediate ones. Note that in (6) the greedy action is used to update the Q function, instead of the value of the actual action taken. To change this behavior, our agent can use other policies to select the action based on the context information and other peculiarties of the problem. In [9] was selected the ε-greedy policy, meaning the agent is greedy most of the time and with a small probability ($\varepsilon > 0$), a random action is selected.

Deep Q-Learning (DQN) with Experience Replay. The main idea is to approximate the Q function (6) using a Neural Network [13]. Its main application is on image-based input, where the Convolutional Neural Networks can approximate the Q function directly from the image pixel. The network receives as input the observation and outputs the value function for each of the available actions. To approximate the Q function as good as possible, the mean square error loss function between the current predicted Q-values (Q_θ) and the true value (Q_{target}) is computed according to:

$$Q_{target}(t) = \begin{cases} r_t, \text{ for terminal } \phi_{t+1} \\ r_t + \gamma \max_{a'}(Q_\theta(\phi_{t+1}, a')), \text{ for non-terminal } \phi_{t+1} \end{cases} \qquad (7a)$$

$$Loss(\theta) = \sum (Q_{target}(t) - Q_\theta(\phi_t, a_t))^2, \qquad (7b)$$

where a corresponds to an action, ϕ to the state and R to the reward. Additionally, t corresponds to the current timestep, Q to the action-value function for each state-action pair, and θ to the network weights. Since the Nerual Network tends to give more importance to the recently seen data during learning and disregard previously found time-based patterns, Experience Replay [8] solves this issue. After completing an episode, it replays the gathered experiences by randomly selecting a batch of a particular size and training the network with it. This replay helps reduce instability produced by training on highly correlated sequential data and increases the learning speed.

2.2   State Representation

Ideally, a notification system would have access to the current activity, emotional state, location, and other private information that we prefer not to utilize to minimize privacy issues. Here, the focus is on using accessible information such as the time or day of the week, the user's reaction to notifications, or the number of notifications already answered. Hence, the aim is to demonstrate that efficient results can be obtained from more simplistic representations of a user's state. Thus, for the DQN and TQ algorithms, the state representations below described were designed.

The elements considered as state information are: (i) Notifications sent and answered during the current epoch, (ii) user's reaction to previous time instant, (iii) time of the day in minutes and (iv) day of the week. State S1 is composed of i–iv, and state S3 of i–iii. These states have a high granularity, that leads to a very large state variable. S2 and S4 have a bigger granularity, replacing minutes for hours. Thus S2 is composed of i–iv in time slots of one hour and S4 is composed of i–iii in slots of one hour.

2.3   Reward Définition

In the case that a notification is sent and the user does not answer, the agent receives reward a. However, if the user responds then the received reward value is b. Contrarily, if a notification is not sent the agent receives c. Lastly, if the episode, in this context a day, ends without having achieved the goal of one answer then d is received. Thus, the rewards assume values in the set $R = \{a, b, c, d\}$. We define the following alternatives for the values of $\{a, b, c, d\}$: R1 = $\{-1, 2, -1, -2\}$; R3 = $\{-2, 2, -1, -3\}$; R5 = $\{-2, 2, 0, -3\}$; R6 = $\{-3, 2, 0, -3\}$.

2.4   Environment Model - Simulation

In our simulated user, we assume that no difference exists between ignoring a message or explicitly dismissing it, considering both as "No Answer". In this way, the cause of not availability (e.g. busy with other activity) is not considered. Furthermore, the user's answer is considered to be either immediate or non-existent.

Behavior Model. This model emulatates a plausible answer by a real user by computing the probability of answering given the sensor data from his/her mobile phone. In the work of [9], the ExtraSensory dataset [15] serves as proxy to the computation of the probability. For each activity label it is defined an ad-hoc value of $P(A|L)$, where A corresponds to answering a notification and L corresponds to the activity labels. Thus, L is a 51-dimensional vector with value 1 in the dimension associated to the current activity. The final response probability model is as follows:

$$P(A \mid L_0, ..., L_i, \overline{L_{i+1}}, ..., \overline{L_{l_t}}) \propto \beta \left(P(A) \prod_{j=0}^{i} P(L_j \mid A) \prod_{k=i+1}^{l_t} P(\overline{L_k} \mid A) \right) \quad (8)$$

$$P(\overline{A} \mid L_0, ..., L_i, \overline{L_{i+1}}, ..., \overline{L_{l_t}}) \propto (1 - \beta) \left(P(\overline{A}) \prod_{j=0}^{i} P(L_j \mid \overline{A}) \prod_{k=i+1}^{l_t} P(\overline{L_k} \mid \overline{A}) \right) \quad (9)$$

For each instance, the above presented factors are estimated, normalized, and, resorting to a simple sampling method, the simulator's response is determined. A random number x that follows $X \sim U(0, 1)$ is sampled. If $P(A|L) > x$ the simulator answers and if $P(\overline{A}|L) > x$ the simulator does not answer.

3 Experimental Description and Metrics

3.1 Model Initialization Methods

One of the main objectives of this paper is to analyze the efficiency levels that algorithms can obtain when models are initialized in different manners. **No Previous Knowledge Models (Online Learning)** - The models start with no prior knowledge, learning only from interaction with a specific user. We expect that this model adapts better to the user, taking longer to reach the better customization. **Previously Trained Models (Offline Learning)** - Here, models are trained with two different users before being tested with a third one, where they only apply what they have learned from previous experience. This approach should reach acceptable results right away. However, the method will not adapt to user input. **Previously Trained Adaptive Models (Combination of Offline and Online Learning)** - In this case, models are likewise trained before being deployed. However, they continue learning, which allows them to start more efficiently than models with no previous knowledge while also growing to be customizable. Assuming the chosen users' routines are varied enough to provide generalized knowledge that could then be applied to any user, this model, which combines the two previous ones, is expected to offer the best and most stable results.

3.2 Daily vs. Weekly Routine

By applying the different state representations of Q-learning and DQN and the different formulations of UCB, this work intends to test if higher levels of efficiency can be obtained when letting the agent learn what a typical week is for the user instead of a typical day. It is expected that when modeling opportune timings throughout a week, the agent takes longer to learn, but if enough time is provided, better results can be obtained.

3.3 Performance Metrics

As performance metrics of our algorithms, we selected two: Goal Achievement Rate (G_r) and Notification Volume (N_v).

$$G_r = \frac{N_A}{N_{Days}} \qquad (10)$$

$$N_v = \frac{N_{Sent}}{N_{Days}} = \frac{\sum_{i=0}^{Days}(N_{A_i} + N_{\overline{A_i}})}{N_{Days}} \qquad (11)$$

G_r, in (10), is the fraction of accepted notifications (N_A) over the number of episodes being tested $(N_{Days}$, each episode representing a day). High G_r values show that our agent was able to identify when users are open to receiving and answering notifications throughout the day. However, an agent may increase the G_r by simply increasing interaction with users. Thus, the volume of sent notifications in (11) is also tracked to balance this effect. A well-behaved agent should have a high response rate (G_r) while maintaining a low notification volume (N_v), ensuring in this way that our system gets a response without bothering the user when he is not receptive.

3.4 Implementation of the Smart Notification App

We implemented a prototype application on Android, where the user answers a set of questions about motivational issues and physical activity. The application implements the UCB Day algorithm, responsible for sending notifications (i.e. actions) of a new questionnaire to be filled to the user. The application of each user is connected to a server that associates questionnaires with users and stores their responses. Since the UCB Day algorithm is personal and its learning process differs for every user, it is running on the personal smartphone of the User.

Algorithm 1. Supporting Functions.

Function AlarmManager():
```
    /* Responsible to call BroadcastReceiver() for every hour
       in the day                                              */
```
while *True -> Every hour* **do**
   | Call **BroadcastReceiver**;
end

Function NewQuizTrigger():
```
    QUIZ_ANSWERS_HOUR = -1; /* Acknowledgement of quiz
        answered hour. Relevant for BroadcastReceiver()      */
    QUIZ_AVAILABLE = True; /* Variable responsible to
        acknowledge if a quiz is available                   */
    FIRST_HOUR_NOTIFICATION = CURRENT_HOUR; /* Acknowledgement
        of quiz received hour. Relevant for
        BroadcastReceiver()                                  */
```
   Call **BroadcastReceiver**;

Function QuizResponse():
```
    QUIZ_ANSWERS_HOUR = CURRENT_HOUR; updateAnsweredHourRewards();
        /* Update Reward B to the rewards array for the
        current hour                                         */
    QUIZ_AVAILABLE = False;
    FIRST_HOUR_NOTIFICATION = -1;
```

The implementation on the application can be followed in the pseudo code of Algorithm 1 and 2. On every episode, a new quiz is associated to an user and immediately, a Firebase [6] Push Notification is sent to the smartphone associated with that same user. This is responsible for triggering $NewQuizTrigger()$, a function that updates stored variables, responsible of letting the system know that a new quiz is available, and used to optimise the $BroadcasteReceiver()$ function calls. The function $NewQuizTrigger()$ also calls $BroadcastReceiver()$ that will run the Upper Confidence Bound (UCB) stated on Sect. 2.1 calculations for the first time this episode, since a Quiz is now available. This function checks if a quiz is available and updates the stored variables responsible for future calculations of the UCB. This function is also responsible for updating the stored reward arrays if the episode ends without an answer from the user, in our case at midnight. After updating the stored variables and arrays, this function calculates the UCB and a decision is made (to send or not to send a notification). When a quiz is answered, the rewards and stored variables are also updated.

The $BroadcastReceiver()$ function, besides being called when a new quiz is available, is also executed every hour of the day, for an ideal number of 24 runs a day.

Algorithm 2. BroadcastReceiver.

```
Function BroadcastReceiver():
    if QUIZ_AVAILABLE then
        if CURRENT_HOUR =! FIRST_HOUR_NOTIFICATION then
            if CURRENT_HOUR != 0 then
                if QUIZ_ANSWER_HOUR != CURRENT_HOUR - 1 then
                    updatePreviousHour(); /* Update Reward A to the
                        rewards array of the previous hour        */
                end
            end
            else
                /* Episode ended without the quiz answered.
                   Midnight update with reward D                 */
                updateHour23();
                updatePreviousHours();
            end
        end
    end
    SEND_NOTIFICATION = calculateUCB(); /* UCB Day algortihm call.
        Will decide if a notification is sent or not.            */
    if SEND_NOTIFICATION then
        /* Updates Notification Array on the current hour         */
        updateNotificationSent();
        sendNotification();
    end
    else
        updateNotificationNotSent();
    end
```

---

4 Results and Discussion

Previous work [9] performed several experiments in simulation, considering all the possible combinations of initialization algorithms, states and rewards. We start by showing the most promising ones from simulation, to select the algorithm that was implemented in an app and used by real life users. All graphs presented show the average result among tested users, employing the N_v and G_r metrics. The standard deviation was also analyzed and is likewise depicted in the displayed graphs. Furthermore, in the tables shown throughout this section, the average G_r and N_v values obtained over 300 days of training are presented.

4.1 Simulation Results

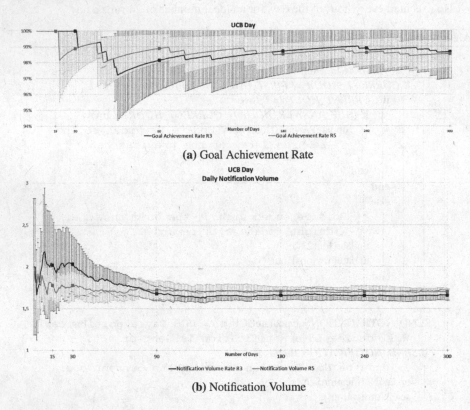

(a) Goal Achievement Rate

(b) Notification Volume

Fig. 1. ComboA: Previously Trained Adaptive, UCB Day - average result over users and across 300 days of training. Source: [9].

Best Combination. As previously presented in [9], the (multiple one hour time slots) UCB Day and reward R5 provided the best result, for both types of agent initialization. By having 24 UCB instances it is possible to have a stateless algorithm that learns to notify timely. Figure 1 shows the results for the Previously Trained Adaptive, UCB Day and reward R5, which provides a high high G_r rate from the model's deployment, while being able to adapt over time to the user's specific routine, achieving a lower N_v as time goes on. In contrast, R6 produced a worse user experience due to the higher penalization value for unanswered notifications, leading to lower N_v and lower G_r.

Figure 2 shows the results of the No Previous Knowledge method, also resorting to UCB Day and reward R5. It is clear that the algorithm needs a longer time to reach similar performances as the Previously Trained initialization. It takes approximately two months to achieve G_r values equivalent to the ones obtained with the initially discussed combination. The promising results with UCB day and reward R5 led us to select this combination for the implementation of the app with real world users.

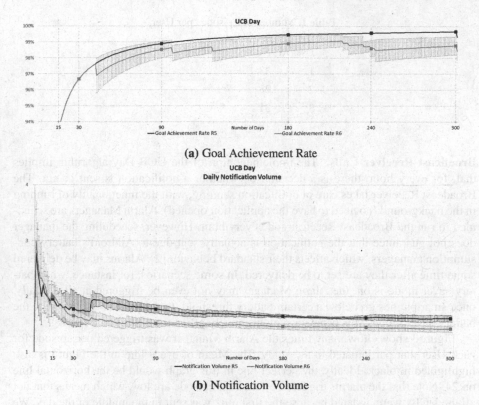

(a) Goal Achievement Rate

(b) Notification Volume

Fig. 2. ComboB: No Previous Knowledge, UCB Day - average result among tested users over 300 days of training. Source: [9].

Best State Representation. S1 and S3 tend to perform better throughout all combinations, revealing that more complex, detailed states are not necessarily always more efficient. However, it is not clear if S3 (average day) is better than S1 (average week).

Best Reward Structure. The reward which generated a better overall performance in the tackled problem was R5. These values provided the best balance between achieving one daily answered notification and not bothering the user.

4.2 Real-World Pilot

We asked 4 volunteers to use the application on their smartphones. The UCB Day implemented in the App run on their phones during several days. Each application is connected to a Server that associates quizzes with the users and stores their answers. Since the UCB Day algorithm is personal and its learning process differs for every user, it is running on the user's personal smartphone. The number of episodes of each user is different, as Table 1 shows since users started this experience at different times.

Table 1. Number of Episodes per User.

User Id	Number of episodes
1	33
2	28
3	31
4	18

Broadcast Receiver Calls. The implementation of the UCB Day algorithm implies that, for every hour, there is a decision on whether a notification is sent or not. The Broadcast Receiver takes care of notification sending, with the functionality of running in the background (no need to have the application opened). Alarm Managers are scheduled to run the Broadcast Receiver code every hour. However, scheduling the manager does not guarantee that the notification is actually sent due to Android's battery consumption managers, which affects their standard behavior [2]. Alarms may be delivered some time after they are set to be delivered. In some scenarios, for instance, when battery saver mode is on, the Alarm Manager may not even be triggered at all. Usually, once smartphones go below a certain battery threshold, they automatically activate the battery-saver mode. This behavior may negatively affect the algorithm's results.

Figure 3 shows how many times the Alarm Manager was triggered per episode for each User that participated in the study. The Mean of every line in the Figure is also highlighted in black. Ideally, the perfect line in this graph would be the horizontal line on 24. Note that the alarms triggered on the first episode are low, which means that not all the UCBs were updated because the first quiz was sent in the middle of the day. We observe that on average, the various instances of the UCB were updated during each day.

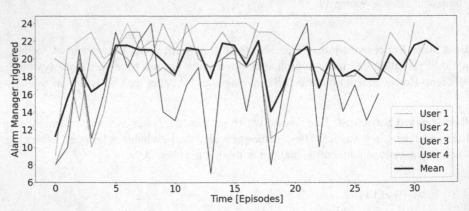

Fig. 3. Number of times Alarm Manager was triggered per Episode.

Suppose the Android Operative System decides not to trigger the Alarm Manager during the most opportune hour for the User to answer quizzes. In that case, no notification will be sent and the overall performance of the algorithm will be affected. However, not triggering the Alarm Manager may not even be prejudicial at all, e.g. for hours after answering a quiz for a given episode.

Notification Statistics. In order to analyze the performance of the Real-world pilot, a set of metrics were logged throughout this experience for every user. Understanding how many notifications are sent, ignored and accepted throughout time is essential.

Figures 4, 5 and 6 show the number of notifications sent (N_{Sent}), ignored ($N_{\overline{A}}$) and accepted (N_A) overtime per episode, respectively. Finally, Fig. 7 showcases the mean of the previous Figures on the same graph, so it is possible to relate these three metrics.

The first evident conclusion is the initial Learning curve every graph follows - it is necessary to send many notifications in the first few days to understand the most opportune hour for the user to accept the notification.

It is also easily identifiable that each learning curve after the initial one originated from a period in which a user ignored several notifications. Therefore, the algorithm needs to relearn the most opportune hour for the user to accept the notification. This behavior is visible with User 3. User 3 does not answer the quiz on Episode 6, and it causes the algorithm to relearn the best hour in the following episodes. When User 3 finally answers the Quiz in Episode 10, the algorithm stabilizes since it learned a new opportune hour, which was kept in the following days.

Ideally, this experiment should last longer to understand more clearly how these curves vary over a more extended period of time. However, for a higher number of episodes, each line tends to the expected behavior: one notification answered per day, with the least possible notifications ignored (ideally, zero).

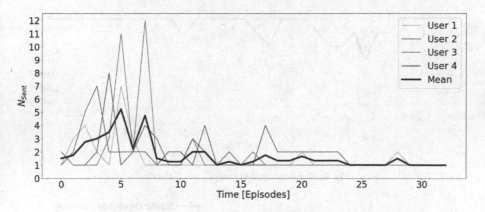

Fig. 4. Notifications Sent per Episode.

Table 2. Goal Achievement Rate and Notification Volume per User.

User	G_r	N_v
1	0.939	1.485
2	0.929	1.893
3	0.742	1.903
4	0.667	2.278

Overall Results and Discussion. Table 2 showcases the Goal Achievement Rate in Eq. (10) and Notification Volume in Eq. (11), of all the users of this experiment. The average and standard deviation are as follows:

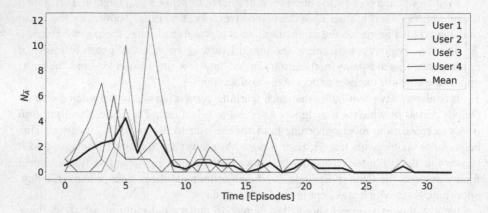

Fig. 5. Notifications Ignored per Episode.

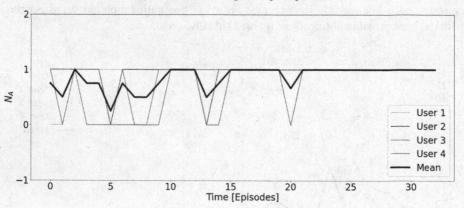

Fig. 6. Notifications Answered per Episode.

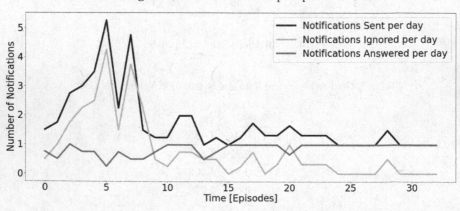

Fig. 7. Relation between the Mean of Notifications Sent, Ignored and Accepted.

1. Goal Achievement Rate: 0.819 ± 0.136
2. Notification Volume: 1.890 ± 0.324

Although these results are worse than the ones presented in Sect. 4.1, one has to consider the number of episodes considered in both scenarios. Whilst the Simulation analysis considered 300 episodes, the real-life pilot was conducted with slightly over 30 episodes. This conclusion may be evident in the fact that the standard deviations of the previous values are so significant.

5 Conclusions and Future Work

This work aims to develop and implement a pilot study for an intelligent notification system that finds the most appropriate time of the day to ask the user to fill out a well-being questionnaire, which acts as a mHealth self-reporting application. In a simulated environment, we studied a set of RL algorithms, to determine the most desirable combination of initialization method, algorithm, state, and reward definition. From the simulated user results the most promising algorithm is UCB-Day, which considers multiple instances of UCB (one per hour). We selected this approach and implement an android application that runs the UCB for each user.

The application was deployed on a pilot with four users that run for around a month. The application is able to reduce the number of needed notifications over time, while being able to save the answers of the questionnaire in a database. Future work should run a pilot with more users with randomized samples on lifestyles, contexts and demographics. In addition, run the algorithm for a longer period of time to evaluate better the adaptation to new routines. Finally, we should consider more elaborated answers in the implemented application, for example, introducing oblivious dismissal (notification goes unnoticed) and intentional dismissal (people decide not to address it).

Acknowledgements. This work has been partially funded by the project LARSyS - FCT Project UIDB/50009/2020 and the project and by the project IntelligentCare - Intelligent Multimorbidity Managment System (Reference LISBOA-01-0247-FEDER-045948), which is co-financed by the ERDF - European Regional Develpment Fund through the Lisbon Portugal Regional Operational Program - LISBOA 2020 and by the Portuguese Foundation for Science and Technology - FCT under CMU Portugal.

References

1. Bellman, R.: Dynamic Programming, 1st edn. Princeton University Press, Princeton (1957)
2. Developers, A.: Manager, alarm (2022). https://developer.android.com/reference/android/app/AlarmManager.html. Accessed 14 Nov 2022
3. Fiordelli, M., Diviani, N., Schulz, P.J.: Mapping mhealth research: a decade of evolution. JMIR **15**(5), 1–14 (2013). https://doi.org/10.2196/jmir.2430
4. Fukazawa, Y., Yamamoto, N., Hamatani, T., Ochiai, K., Uchiyama, A., Ohta, K.: Smartphone-based mental state estimation: a survey from a machine learning perspective. JIP **28**, 16–30 (2020). https://doi.org/10.2197/ipsjjip.28.16
5. Geneva: WHO: noncommunicable diseases progress monitor 2020 (2020). https://doi.org/10.5005/jp/books/11410_18

6. Google: Firebase (2022). https://firebase.google.com/. Accessed 14 Nov 2022
7. Ho, B.J., Balaji, B., Koseoglu, M., Srivastava, M.: Nurture: notifying users at the right time using reinforcement learning. In: UBICOMP, pp. 1194–1201 (2018). https://doi.org/10.1145/3267305.3274107
8. Lin, L.J.: Reinforcement learning for robots using neural networks. Ph.D. thesis (1993)
9. Machado., A., Cardoso., H., Moreno., P., Bernardino., A.: Active data collection of health data in mobile devices. In: Proceedings of the 3rd International Conference on Deep Learning Theory and Applications - DeLTA, pp. 160–167. INSTICC, SciTePress (2022). https://doi.org/10.5220/0011300700003277
10. Mehrotra, A., Musolesi, M.: Intelligent notification systems: a survey of the state of art and research. Challenges 1(1), 1–26 (2017)
11. Mehrotra, A., Musolesi, M., Hendley, R., Pejovic, V.: Designing content-driven intelligent notification mechanisms for mobile applications. In: UBICOMP, pp. 813–824 (2015). https://doi.org/10.1145/2750858.2807544
12. Mehrotra, A., Pejovic, V., Vermeulen, J., Hendley, R.: My phone and me: understanding people's receptivity to mobile notifications. In: CHI, pp. 1021–1032 (2016). https://doi.org/10.1145/2858036.2858566
13. Mnih, V., Kavukcuoglu, K.: Silver: human-level control through deep reinforcement learning. Nature 518(7540), 529–533 (2015)
14. Richard S. Sutton, A.G.B.: Reinforcement Learning : an Introduction. MIT Press, Cambridge 2° edn. (2018). https://doi.org/10.5555/3312046
15. Vaizman, Y., Ellis, K., Lanckriet, G.: Recognizing detailed human context in the wild from smartphones and smartwatches. IEEE Pervasive Comput. 16(4), 62–74 (2017). https://doi.org/10.1109/MPRV.2017.3971131
16. Vishwanath, S., Vaidya, K., Nawal, R.: Touching lives through mobile health-assessment of the global market opportunity. In: PwC (2012)
17. Watkins, P.: Q-learning. Mach. Learn. 8(3–4), 279–292 (1992). https://doi.org/10.1109/ICCC49849.2020.9238991
18. Whitmore, C., et al.: The relationship between multimorbidity and self-reported health among community-dwelling older adults and the factors that shape this relationship: a mixed methods study protocol using clsa baseline data. Int. J. Qual. Methods 20, 16094069211050166 (2021)
19. WHO Global Observatory for eHealth: mHealth: new horizons for health through mobile technologies: second global survey on eHealth (2011)

Reliable Classification of Images by Calculating Their Credibility Using a Layer-Wise Activation Cluster Analysis of CNNs

Daniel Lehmann[✉] and Marc Ebner

Institut für Mathematik und Informatik, Universität Greifswald, Walther-Rathenau-Straße 47, 17489 Greifswald, Germany
{daniel.lehmann,marc.ebner}@uni-greifswald.de

Abstract. An image classification model based on a Convolutional Neural Network architecture generally achieves a high classification performance over a wide range of image domains. However, the model is only able to obtain such a high performance on in-distribution samples. On out-of-distribution samples, in contrast, the performance of the model may be significantly decreased. To detect out-of-distribution samples, Papernot and McDaniel [38] introduced a method named DkNN, which is based on calculating a sample credibility score by a nearest neighbor classification in feature space of the hidden layers of the model. However, a nearest neighbor classification is memory-intensive and slow at inference. To address these issues, Lehmann and Ebner [26] suggested a method named LACA, which calculates the credibility score based on clustering instead of a nearest neighbor classification. Lehmann and Ebner [26] showed that for out-of-distribution samples with respect to models trained on MNIST, SVHN, or CIFAR-10, LACA is significantly faster at inference compared to DkNN, while obtaining a similar performance. In this work, we conducted additional experiments to test LACA on more complex datasets (Imagenette, Imagewoof). Our experiments show that LACA is significantly faster at inference compared to DkNN also for these more complex datasets. Furthermore, LACA computes meaningful credibility scores, while DkNN fails on these datasets.

Keywords: CNN · Out-of-Distribution Detection · Clustering

1 Introduction

A classification model based on a Convolutional Neural Network (CNN) architecture is the preferred approach for classifying images due to the high classification performance of such a model over a wide range of image domains [12,20]. We usually train such a model using a training dataset until the model achieves a sufficient classification performance on a testing set. Then, we may deploy the model to a production system to classify real-world images. However, the classification performance of the model may be significantly decreased for the real-world images compared to the images of the testing set. A CNN-based model only achieves a high classification performance on in-distribution samples. In-distribution samples are samples that were drawn from the training data distribution of the model. The real-world images from the production

A. Fred et al. (Eds.): DeLTA 2022, CCIS 1858, pp. 33–55, 2023.
https://doi.org/10.1007/978-3-031-37317-6_3

system, however, may be out-of-distribution samples. Out-of-distribution samples were drawn from a data distribution that is different from the training data distribution. The model has not learned anything about these samples during model training. Thus, the classification performance of a CNN-based model is usually low for out-of-distribution samples [26]. Out-of-distribution samples can appear in the form of images showing an object of a class the model has not seen during training, or images showing an object of a known class but the model has seen this object in a different situation during training. A CNN-based model trained to classify images of various types of whole apples, for instance, will probably fail to predict the correct class for an image showing slices of those apples. Furthermore, the model will not be able to predict the correct class for an image showing a different object such as an orange. Nevertheless, the model may incorrectly classify that image as a certain type of apple with high confidence. We refer to these kinds of out-of-distribution samples as natural out-of-distribution samples [15]. However, out-of-distribution samples do not only occur naturally but they can also be generated artificially by an attacker from in-distribution samples. This kind of out-of-distribution sample is typically referred to as an adversarial sample [1,9,44]. However, CNN-based models not only fail to predict the correct class for an out-of-distribution sample, but these models also fail without any warning. The model may predict an incorrect class despite achieving a high softmax score for the predicted class [7,13]. As a consequence, using CNN-based models for safety-critical applications is challenging (e.g., for driving assistance systems, or medical diagnosis systems).

To improve the reliability of CNN-based models, a wide range of methods have been proposed to detect out-of-distribution samples. A promising method was proposed by Papernot and McDaniel [38]: *Deep k-Nearest Neighbors* (DkNN). The DkNN method calculates a credibility score for a sample. This credibility score expresses the likelihood of the sample being drawn from the training data distribution. Thus, if the score is high, the sample is most likely an in-distribution sample. If the score is low, however, the sample is most likely an out-of-distribution sample. The DkNN method is based on the assumption that an in-distribution sample of a specific class is always located near other in-distribution samples (e.g., training samples) of the same class in feature space of each hidden layer of a CNN-based model. Based on this assumption, Papernot and McDaniel [38] calculate the credibility of a sample by performing a k-nearest neighbor classification in feature space of the hidden layers of the model (i.e., on the hidden layer activations). If the k-nearest neighbor classifiers predict the same class for the sample at all hidden layers, the sample will most likely be an in-distribution sample. If the k-nearest neighbor classifiers predict different classes for the sample across the different hidden layers, on the other hand, the sample may be an out-of-distribution sample. However, a k-nearest neighbor classification is typically slow at inference because the sample needs to be compared to every training sample. Moreover, the DkNN method applies the k-nearest neighbor classification not only once but for every hidden layer of the model. Hence, to decrease the runtime of their method, Papernot and McDaniel [38] used an approximate k-nearest neighbor classification. An approximate k-nearest neighbor classification compares the sample only to a subset of the training data samples at inference. However, this subset usually still contains a high number of training data samples. Hence, using an approximate k-nearest neighbor classification improves the runtime of the DkNN method at inference, but the method is still relatively slow. However, at inference, we usually want to find out quickly whether a sample is out-of-distribution. Moreover, the k-nearest neighbor classification also requires storing the whole training set for inference. This may become challenging for training sets containing a huge number of samples. To address the issues of the DkNN method, Lehmann and Ebner [26] proposed a method based on a *layer-wise activation cluster analysis* (LACA) for calculating the sample credibility (described in Sect. 3). The LACA method calculates the credibility score of a sample by using clustering at the hidden layers of the model instead of a k-nearest neighbor classification. Lehmann and Ebner [26] showed that the LACA method is significantly faster than the DkNN method while achieving similar performance. Moreover, the LACA method does not require storing the whole training set for inference.

However, in [26] the LACA method was only tested on simple datasets. The focus of [26] was not to present a ready-to-use out-of-distribution detection method yet but to take the first step and examine whether the LACA method is able to calculate meaningful credibility scores at all, and how the LACA method performs in comparison to the DkNN method suggested by Papernot and McDaniel [38], who also used these simple datasets in their experiments. Furthermore, it has not been examined yet whether all hidden layers of a model are necessary for the credibility calculation. Thus, in this work, we take the next step and state the following two research questions: (1) Can we further decrease the runtime of the LACA method by omitting some of the hidden layers for the credibility calculation without significantly lowering its performance?, and (2) Does the LACA method also work for more complex datasets in a reasonable time compared to the DkNN method? To answer these questions, we conducted several additional experiments (Sect. 4) to the experiments from Lehmann and Ebner [26]. Our contributions are as follows: (1) We examined whether the runtime of the LACA method can be further improved by excluding some of the lower hidden layers from the credibility calculation (Sect. 4.2), and (2) we tested the LACA and the DkNN method on more complex images that are more likely to occur in practice (Sect. 4.3).

2 Related Work

A vast number of methods have been proposed for detecting out-of-distribution samples. Popular methods include Bayesian Neural Networks [8], detecting out-of-distribution samples by an additional model output [10], an out-of-distribution detector based on a generative model [34], detecting out-of-distribution samples by adjusting the loss function [23], or using self-supervised learning to detect out-of-distribution samples [14].

However, similar to the LACA method, there have also been studies that propose to detect out-of-distribution samples by exploiting the hidden layer activations of a model. Sastry and Oore [42] proposed a method based on analyzing hidden layer activations by using Gram matrices for detecting out-of-distribution samples. Li and Li [28], on the other hand, used convolutional filter statistics for a cascade-based out-of-distribution detector. Ma et. al. [30] obtained local intrinsic dimensionality estimates from the hidden layer activations. They showed that these estimates can be used to detect out-of-distribution samples. Carrara et. al. [2] suggested an out-of-distribution detection method based on a k-nearest neighbor scoring on the hidden layer activations. Lee et. al. [24] suggested a method for calculating the confidence of a model prediction based on information obtained from class-conditional Gaussian distributions fitted at each hidden layer of the model. Chen et. al. [4] obtained the confidence of the model prediction, on the other hand, by a meta-model trained on the hidden layer activations of the model. Cohen et. al. [6] introduced a method for detecting out-of-distribution samples that uses sample influence scores in combination with a k-nearest neighbor classification applied on the hidden layer activations. Lin et. al. [29] proposed a multi-level approach to detect out-of-distribution samples. Metzen et. al. [35] suggested an out-of-distribution detector in form of a subnetwork attached to a particular hidden layer of the model. However, none of these methods exploit cluster information from the hidden layer activations. Huang et. al. [16] proposed an out-of-distribution detection method based on such cluster information. They observed that out-of-distribution samples cluster together in feature space. However, in contrast to the LACA method, they detect if a sample is out-of-distribution by checking whether the sample is located within a certain distance to the center of the out-of-distribution cluster in feature space.

Cluster information from the activations of the hidden layers of a CNN-based model has also been used in several other studies. Nguyen et. al. [37] identified clusters at the last hidden layer of a trained model to visualize the multifaceted features that were learned by the model. Lehmann and Ebner [27] used the cluster information at a higher hidden layer for undersampling an imbalanced training dataset. Chen et. al. [3] used the cluster information at the final hidden layer, on the other hand, to detect whether the training dataset of the model was poisoned.

3   Method

We are given a CNN-based model f that was trained to classify images into one of C categories. After model training, f should be used to predict the correct class of an image x^I at inference. However, if image x^I is an out-of-distribution sample, our model f will most likely fail to make the correct class prediction for x^I. To be able to detect whether x^I is out-of-distribution, a method named *Layer-wise Activation Cluster Analysis* (LACA) was introduced in [26]. LACA calculates a credibility score $credib(x^I) \in [0, 1]$ for image x^I at inference. This credibility score $credib(x^I)$ expresses the likelihood of x^I being drawn from the same data distribution as the images that were used for training the model f. Thus, if x^I is an out-of-distribution sample, $credib(x^I)$ will be low. If x^I is an in-distribution sample, however, $credib(x^I)$ will be high. Calculating the credibility score $credib(x^I)$ of an image x^I at inference (Sect. 3.3) requires several in-distribution statistics. These statistics can be computed before inference (Sect. 3.2). The approach for obtaining the in-distribution statistics is based on finding clusters in the hidden layer activations of model f as described in Sect. 3.1.

3.1   Identifying Clusters in Layer Activations

Obtaining the in-distribution statistics necessary for calculating the credibility $credib(x^I)$ of image x^I at inference with respect to model f (containing L layers) is based on finding clusters in the activations of the hidden layers $l \in 2, .., L - 1$ of f, i.e., the layers between the input layer containing the raw image data and the output layer containing the softmax scores. The activations of these hidden layers represent the intermediate representations of the images fed into model f towards the classification objective of separating these images by their classes. We choose one of these hidden layers l to explain how clusters can be identified in the activations of that layer.

First, we feed all N training images x^D into model f again. However, we do not feed these images into f for further training but to receive the activations of the training images at hidden layer l. The weights of f are kept fixed. After feeding a training image x^D into model f, we fetch the activations of x^D at the chosen layer l. If l is a convolutional layer, we receive the fetched activations in shape of a three-dimensional tensor. To be able to search for clusters in the activations, however, they need to be vector-shaped. Thus, we flatten the three-dimensional tensor to a vector. However, this step is only required for convolutional layers. If l is a linear layer, we receive the fetched activations as a vector already. As a result, we obtain for each training image x^D an activation vector $a^l_{x^D}$ of a layer-specific vector size M_l. After obtaining the activation vector $a^l_{x^D}$ for all N training images x^D, we concatenate these activation vectors to an activation matrix A^l_D of size $N \times M_l$.

In this matrix A^l_D we aim to identify clusters. However, matrix A^l_D is commonly high-dimensional due to the high number of activations M_l received from hidden layer l. Algorithms for searching clusters, on the other hand, generally do not work well in high-dimensional spaces. Thus, we project A^l_D from size $N \times M_l$ down to size $N \times 2$ using dimensionality reduction beforehand. First, as a preprocessing step for the dimensionality reduction, we need to scale the values of matrix A^l_D. From each value a_{ij} of A^l_D we subtract the mean of row i and then divide by the standard deviation of row i. After scaling all values of A^l_D, we reduce the size of A^l_D from $N \times M_l$ down to $N \times 50$ using PCA [39] (linear dimensionality reduction) followed by reducing the size of A^l_D a second time from $N \times 50$ down to $N \times 2$ using UMAP [33] (non-linear dimensionality reduction). In theory, it may be better to project the dimensions of A^l_D from $N \times M_l$ directly down to $N \times 2$ using a non-linear dimensionality reduction technique such as UMAP [33]. In practice, however, applying a non-linear dimensionality reduction on a high-dimensional matrix such as A^l_D is computationally expensive, as pointed out by Lehmann and Ebner [25]. Thus, we need this two-step process for projecting A^l_D to a lower-dimensional space. Lehmann and Ebner [25] showed that combining PCA [39] and UMAP [33] works best for projecting layer activation data from CNN-based models. As a result, we receive the projected matrix $r^l_D(A^l_D)$ of size $N \times 2$ along with the projection model r^l_D (combination of the PCA [39] model and the UMAP [33] model).

After receiving the projected activation matrix $r^l_D(A^l_D)$, we search for clusters in this matrix. Lehmann and Ebner [25] showed that the k-Means algorithm [31] works best for finding clusters in the layer activations from CNN-based models. Parameter k of the k-Means algorithm specifies the number of clusters to be found. Lehmann and Ebner

[26] choose k by the silhouette score [40]. The silhouette score is a cluster quality metric expressing how well clusters are separated based on the average distance between data samples of different clusters (inter-cluster distance) and the average distance between data samples of the same cluster (intra-cluster distance). Chen et. al. [3] showed that the silhouette score works best for evaluating clusters that were identified in the layer activations from CNN-based models. To find parameter k, we check different values for k within a range of $C - 5, .., C + 5$. This range is inspired by the classification objective of f during model training to find a representation of the training images that is linearly separable according to the classes of these images. Thus, we expect to find exactly one cluster for each of the C classes in the activations from the higher hidden layers of f. In the activations from the lower hidden layers, in contrast, the number of clusters is expected to differ slightly. Thus, we consider the range of $C - 5, .., C + 5$ for parameter k. With each value of this range, we search for clusters in $r_D^l(A_D^l)$ using k-Means and receive a clustering result each time. The value for k that yields the clustering result with the best silhouette score is chosen. Finally, we receive H^l ($H \in C - 5, .., C + 5$) clusters along with the clustering model g_D^l for hidden layer l.

3.2   Obtaining In-Distribution Statistics

Calculating the credibility score $credib(x^I)$ of an image x^I at inference (with respect to model f containing L layers) requires several in-distribution statistics. These in-distribution statistics can be computed before inference. One type of these statistics is the class-distribution statistic, which we obtain from the activations of the training samples (x^D, y^D) of model f at the hidden layers $l \in 2, .., L-1$ of f. In order to receive the class-distribution statistic from a specific hidden layer l, we must identify clusters in the layer activations of the training images x^D at l beforehand. To identify the clusters, we use the approach that was described in Sect. 3.1. As a result, we obtain H^l clusters from the layer activations of l. Each identified cluster $h^l \in 1, .., H^l$ contains a subset $(x_{h^l}^D, y_{h^l}^D)$ of the training samples (x^D, y^D) in feature space of l (i.e., in form of their layer activations). For each cluster h^l, we check which of the C classes can be found in h^l among the training samples $(x_{h^l}^D, y_{h^l}^D)$ that are part of h^l. Then, we calculate the percental distribution of these classes, i.e., for each class $c \in 1, .., C$: What percentage $p_{h^l}^l(c)$ of the training samples $(x_{h^l}^D, y_{h^l}^D)$ in h^l is of class c? This percental distribution is used as the class-distribution statistic $S_D^l(h^l)$ of cluster h^l (Eq. 1). However, we expect the training dataset to contain outliers. As a consequence, some clusters may contain only a very low number of training samples of a certain class. To avoid such outliers, we consider for the class-distribution statistic $S_D^l(h^l)$ only those classes c in h^l whose percentage $p_{h^l}^l(c)$ is greater than a threshold t (e.g., $t = 0.05$). This threshold t is a parameter of the method and needs to be specified beforehand.

$$S_D^l(h^l) = \left\{ \left(c, p_{h^l}^l(c) \right) \,\middle|\, c \in 1, ..., C \,,\, p_{h^l}^l(c) > t \right\}$$

$$p_{h^l}^l(c) = \frac{|(x_{h^l}^D, y_{h^l, y==c}^D)|}{|(x_{h^l}^D, y_{h^l}^D)|}$$

(1)

The class-distribution statistics of all identified clusters at layer l yield the class-distribution statistic S_D^l of that layer. Finally, we repeat this process for the remaining hidden layers of model f. As a result, we obtain a class-distribution statistic from each hidden layer of f.

In addition to the class-distribution statistics S_D^l for each hidden layer l, we also compute a layer score $w^l \in [0, 1]$ for l. This score reflects how the classes are distributed in the identified clusters at that layer. As pointed out by Zeiler and Fergus [45], the lower hidden layers detect low-level image features (e.g., edges, corners, simple shapes). Low-level image features are not class-specific. Hence, these features can even be shared between different classes. An image of a soccer player and an image of a baseball player, for instance, share specific low-level image features such as simple shape features of the players (e.g., from the face or the sportswear) and the playing field (e.g., from the grass). As a consequence, images of different classes can be close to each other in feature space at lower hidden layers (as shown in Fig. 1). Thus, the clusters at the lower hidden layers of f usually contain a high number of classes, which tend to be rather uniformly distributed. This kind of class distribution results in a low layer score w^l. The higher hidden layers, in contrast, detect high-level image features as shown by Zeiler and Fergus [45] (e.g., complex shapes, object parts, objects). High-level image features are rather class-specific. This follows from the classification objective of the model. During training, the model tries to incrementally find a representation of the training images that is linearly separable according to their classes. Thus, when going from the first hidden layer to the last hidden layer of the model, images of the same class are pushed increasingly closer together, while images of different classes are pushed increasingly farther apart from each other in feature space of the layers (as shown in Fig. 1). Thus, the clusters at the higher hidden layers usually contain only a few classes, which tend to become increasingly imbalanced. At the final hidden layer, the clusters generally even contain a majority class with a frequency of at least 90%. This kind of class distribution results in a high layer score w^l. To compute the layer score w^l for a hidden layer l, Lehmann and Ebner [26] suggested a very simple approach. For each found cluster h^l at l, we select the class with the second-highest and the class with the highest percental occurrence from the class distribution statistic $S_D^l(h^l)$ of that cluster and compute the absolute difference of their percental occurrence values. As a result, we receive a score $w_{h^l}^l$ (the absolute difference value) from each cluster h^l of l. To obtain the unnormalized layer score w_U^l of l, we average over the scores $w_{h^l}^l$ from all clusters of l. Finally, we normalize the scores w_U^l from all layers to receive the normalized layer scores $w^l \in [0, 1]$, $\sum_l w^l = 1$.

After receiving a class-distribution statistic S_D^l at each hidden layer l of f, we also need to compute a cluster-distribution statistic at the hidden layers. However, we do not obtain these cluster-distribution statistics from the samples (x^D, y^D) of the training set of f but from the samples (x^T, y^T) of a different dataset (e.g., a held-out test set). The samples (x^T, y^T) of this dataset were also sampled from the same data distribution as the training samples (x^D, y^D), but model f did not see any of these samples (x^T, y^T) during training. Similar to the credibility calculation suggested by Papernot and McDaniel [38], we use this dataset for calibrating the credibility scores at inference (Sect. 3.3). Hence, we refer to the samples (x^T, y^T) as calibration samples in

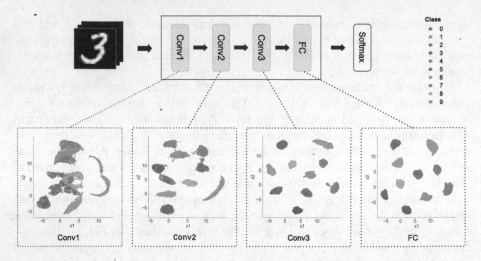

Fig. 1. Visualization of the MNIST training samples in feature space at the hidden layers of the CNN-based model [26].

the following. Calculating the cluster-distribution statistics from the calibration samples (x^T, y^T) is similar to calculating the class-distribution statistics from the training samples (x^D, y^D). First, we create the activation vectors $a^l_{x^T}$ from the calibration images x^T in the same way as the activation vectors $a^l_{x^D}$ from the training images x^D as described in Sect. 3.1. However, we learn neither a projection model nor a clustering model from these activations (as we did from the activations of the training images). Instead, we project each activation vector $a^l_{x^T}$ using the projection model r^l_D, which was learned from the activations of the training images as described in Sect. 3.1. As a result, we receive the projected activation vector $r^l_D(a^l_{x^T})$ for each calibration image x^T. Then, we apply the clustering model g^l_D (learned from the training images as described in Sect. 3.1) on each vector $r^l_D(a^l_{x^T})$. As a result, we obtain for each calibration sample (x^T, y^T) the cluster $h^l_{x^T}$ at hidden layer l in which (x^T, y^T) falls. Finally, we calculate the percental distribution of the classes of the calibration samples (x^T, y^T) over the clusters h^l, i.e., for each cluster h^l at l: What percentage $p^l_c(h^l)$ of the calibration samples $(x^T, y^T_{y==c})$ of class c is located in cluster h^l? This percental distribution is used as the cluster-distribution statistic $S^l_T(c)$ of class c (Eq. 2).

$$S^l_T(c) = \left\{ \left(h^l, p^l_c(h^l) \right) \;\middle|\; h^l \in 1, ..., H^l \right\}$$

$$p^l_c(h^l) = \frac{|(x^T_{h^l}, y^T_{h^l, y==c})|}{|(x^T, y^T_{y==c})|}$$

(2)

The cluster-distribution statistics of all classes c at hidden layer l yield the cluster-distribution statistic S^l_T of that layer. Finally, we repeat this process for the remaining hidden layers of model f. As a result, we obtain a cluster-distribution statistic from each hidden layer of f.

3.3 Calculating the Credibility of an Image at Inference

The credibility score $credib(x^I) \in [0,1]$ of an image x^I at inference expresses the likelihood that x^I is an in-distribution sample. To calculate $credib(x^I)$, we use the in-distribution statistics that were obtained before inference (Sect. 3.2). Our assumption about in-distribution samples states that an in-distribution sample of a specific class is always located near other in-distribution samples (e.g., training samples) of the same class in feature space of each hidden layer of the model. The true class of x^I is unknown. However, we can check the classes y^D of the training samples (x^D, y^D) of model f (in-distribution samples) that are close to x^I in feature space of each hidden layer l of f to see whether our assumption is violated. First, we feed x^I into model f to create the activations of x^I at the hidden layers of f. We obtain the activation vector $a_{x^I}^l$ at each hidden layer l in the same way as we obtained the activation vectors $a_{x^D}^l$ from the training images x^D (as described in Sect. 3.1). Then, we project $a_{x^I}^l$ using the projection model r_D^l, which was learned from the activations of the training images (as described in Sect. 3.1). As a result, we receive the projected activation vector $r_D^l(a_{x^I}^l)$ of x^I. We apply the clustering model g_D^l (learned from the training images as described in Sect. 3.1) on this vector $r_D^l(a_{x^I}^l)$ to obtain the cluster $h_{x^I}^l$ at layer l into which image x^I falls. Through the class-distribution statistic $S_D^l(h_{x^I}^l)$ of $h_{x^I}^l$ (obtained from the training samples as described in Sect. 3.2) we can identify the classes of the training samples that are also in cluster $h_{x^I}^l$. These training samples are the in-distribution samples that are close to x^I. Thus, the classes of these training samples form the set of potential classes $cset_{x^I}^l$ of x^I at hidden layer l. However, according to our assumption about in-distribution samples, the true class of an in-distribution sample is not close to a specific class at one hidden layer but all hidden layers. Hence, we take the intersection of $cset_{x^I}^l$ over all hidden layers l to obtain the final set of potential classes $cset_{x^I}$ of x^I (Eq. 3).

$$cset_{x^I} = \bigcap_l cset_{x^I}^l \tag{3}$$

If $cset_{x^I}$ is empty, however, we can immediately conclude that x^I is probably out-of-distribution (as shown in Fig. 2). The set $cset_{x^I}$ can only be empty if x^I is close to different classes across the different hidden layers. Thus, our assumption is violated, and we set $credib(x^I)$ to 0. If $cset_{x^I}$ is non-empty, however, we assume that the true class of x^I is in $cset_{x^I}$. Thus, for each class c_{x^I} in $cset_{x^I}$, we compute the credibility score $credib(x^I, c_{x^I})$ of x^I assuming that c_{x^I} is the true class of x^I (as shown in Fig. 2). We calculate $credib(x^I, c_{x^I})$ from the probabilities $p_{c_{x^I}}^l(h_{x^I}^l)$ at each hidden layer l of f, which indicate the likelihood that class c_{x^I} occurs in cluster $h_{x^I}^l$ at l. A calibrated form of this probability can be received from the cluster-distribution statistic $S_T^l(c_{x^I})$ computed from the calibration samples (as described in Sect. 3.2). We can obtain $p_{c_{x^I}}^l(h_{x^I}^l)$ through $S_T^l(c_{x^I})(h_{x^I}^l)$. A high probability $p_{c_{x^I}}^l(h_{x^I}^l)$ indicates that in-distribution samples (from the calibration set) of class c_{x^I} mainly occur in cluster $h_{x^I}^l$ at hidden layer l. Assuming that x^I is of class c_{x^I}, we can conclude that x^I satisfies our assumption about in-distribution samples at the layer because x^I is close to a high number of other in-distribution samples of class c_{x^I}. A low probability $p_{c_{x^I}}^l(h_{x^I}^l)$, on the other hand, indicates that the majority of samples of class c_{x^I} occur in a cluster different to $h_{x^I}^l$.

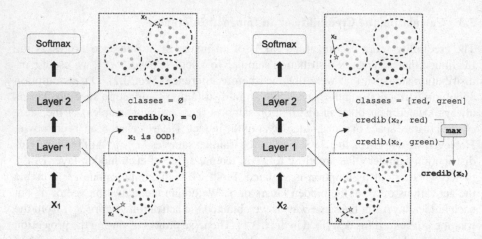

Fig. 2. To calculate the credibility of an image x with respect to a model f, we identify the classes $cset^l$ of the training samples that are located in the same cluster as x at each hidden layer l of f. Then, we take the intersection of $cset^l$ over all hidden layers l to receive the set of common classes $cset$ across the layers. If $cset$ is empty, we conclude that x is an out-of-distribution (OOD) sample (left). Otherwise, we compute the credibility for each class in $cset$, and select the maximum of these credibility values as the final credibility of x (right) [26].

Thus, the samples in cluster $h_{x^I}^l$ of class c_{x^I} might be just outliers at l. Assuming that x^I is of class c_{x^I}, we can conclude that x^I is probably either an outlier as well or it is out-of-distribution because x^I is only close to a few in-distribution samples of class c_{x^I}. However, if $p_{c_{x^I}}^l(h_{x^I}^l)$ is 0 for at least one of the hidden layers l, we can immediately conclude that x^I is probably out-of-distribution because no other in-distribution samples (calibration samples) of class c_{x^I} are in cluster $h_{x^I}^l$ into which image x^I falls. Thus, our assumption is violated, and we set $credib(x^I)$ to 0. If $p_{c_{x^I}}^l(h_{x^I}^l)$ is not 0 at any hidden layer, on the other hand, we use $p_{c_{x^I}}^l(h_{x^I}^l)$ for calculating the credibility $credib(x^I, c_{x^I})$. To calculate the credibility $credib(x^I, c_{x^I})$, we average the probabilities $p_{c_{x^I}}^l(h_{x^I}^l)$ over all hidden layers l. However, as already pointed out in Sect. 3.2, the probabilities $p_{c_{x^I}}^l(h_{x^I}^l)$ of the lower hidden layers are rather low in general as the clusters at these layers contain a high number of classes that tend to be uniformly distributed. The clusters at the higher hidden layers, in contrast, contain only a few classes with one majority class. Thus, a low probability $p_{c_{x^I}}^l(h_{x^I}^l)$ at a higher hidden layer is a stronger indication of x^I being out-of-distribution than a low probability $p_{c_{x^I}}^l(h_{x^I}^l)$ at a lower hidden layer. As a consequence, we take a weighted average of the probabilities $p_{c_{x^I}}^l(h_{x^I}^l)$ over all hidden layers to put more weight on the higher hidden layers than the lower hidden layers. The layer scores w^l that we obtained before inference (as described in Sect. 3.2) are used as the weights. As a result, we receive the credibility score $credib(x^I, c_{x^I})$ for x^I assuming x^I is of class c_{x^I} (Eq. 4).

$$credib(x^I, c_{x^I}) = \sum_l w^l p_{c_{x^I}}^l(h_{x^I}^l) = \sum_l w^l S_T^l(c_{x^I})(h_{x^I}^l) \tag{4}$$

```
def calcCredib(x_i):
    # get potential classes of x_i
    cset = []
    for l in range(1,L):
        a = getActivations(x_i, l)
        h = getCluster(r(l), g(l), a)
        cset_l = getClasses(s_d(l, h))
        cset = intersect(cset, cset_l)

    # check if class set is empty
    if cset is empty:
        return 0  # x_i is out-of-distribution!

    # get credibility score
    credibList = []
    for c in cset:
        for l in range(1,L):
            a = getActivations(x_i, l)
            h = getCluster(r(l), g(l), a)
            prob = getProb(s_t(l, c)(h))
            if prob == 0:
                return 0  # x_i is out-of-distribution!
            credibList.append(prob * w(l))

    return max(credibList)
```

Fig. 3. Python code of the algorithm to calculate the credibility of an image.

To obtain the final credibility score $credib(x^I)$ of x^I, we select the credibility score $credib(x^I, c_{x^I})$ with the highest value over all potential classes c_{x^I} in $cset_{x^I}$ (Eq. 5). The corresponding class is assumed to be the true class of x^I.

$$credib(x^I) = \max_{c_{x^I}} credib(x^I, c_{x^I})$$ (5)

The Python code for the algorithm to calculate the credibility score $credib(x^I)$ of an image x^I is shown in Fig. 3.

4 Experiments

4.1 Baseline

In [26] the LACA method (described in Sect. 3) was tested in comparison to the DkNN method (suggested by Papernot and McDaniel [38]) on three simple datasets: the MNIST dataset (grayscale images, size: 28×28, 10 classes) [22], the SVHN dataset (color images, size: 32×32, 10 classes) [36], and the CIFAR-10 dataset (color images,

Table 1. Accuracies achieved by the MNIST, SVHN, and CIFAR-10 model on the respective in-distribution dataset (InDist), natural out-of-distribution dataset (NatOoD) (MNIST: KMNIST, SVHN: CIFAR-10, CIFAR-10: SVHN), and adversarial datasets (FGSM, BIM, PGD).

Model	InDist	NatOoD	FGSM	BIM	PGD
MNIST	0.9905	0.0759	0.0805	0.0004	0.0246
SVHN	0.8994	0.0924	0.0272	0.0079	0.0242
CIFAR-10	0.9242	0.0935	0.1321	0.0084	0.0093

size: 32×32, 10 classes) [19]. The results obtained in [26] showed that the LACA method is able to achieve similar results as the DkNN method while being significantly faster at inference. In the first experiment, we aimed to reproduce the results obtained in [26].

First, we needed to train a model for each of the three datasets using their respective training sets (MNIST: 60,000 training samples, SVHN: 73,257 training samples, CIFAR-10: 50,000 training samples). For MNIST and SVHN, a simple CNN-based architecture was used. The architecture consisted of three convolutional layers (ConvLayer) and one fully-connected layer (FC): ConvLayer1 (kernel size: 8, stride: 2, filters: 64) - ConvLayer2 (kernel size: 6, stride: 2, filters: 128) - ConvLayer3 (kernel size: 5, stride: 1, filters: 128) - FC (size: 10). The ReLU activation function was applied on the output of each convolutional layer. To initialize the model weights we chose a standard Kaiming Uniform [11] initialization. We trained the MNIST model for 6 epochs with a fixed learning rate of 0.001. The SVHN model, on the other hand, was trained for 18 epochs using a multi-step learning rate (LR) schedule (base LR: 0.001, step at epoch: (10,14,16), gamma: 0.1). Both models were trained using the Adam optimizer [18]. For CIFAR-10, on the other hand, a 20-layer residual network (ResNet) architecture [12] (suggested by Zhang et. al. [46]) was chosen. The Fixup initialization [46] was used to initialize the model weights. To extend the variety of the CIFAR-10 training images, we added several data augmentation techniques to the training setup (random horizontal flip, random crop, mixup). We trained the CIFAR-10 model for 200 epochs with the SGD optimizer using a cosine-annealing learning rate (LR) schedule (base LR: 0.1).

Similar to the DkNN method [38], the LACA method requires a calibration set to obtain certain in-distribution statistics necessary for calculating the credibility of an image (as pointed out in Sect. 3.2). We set the size of this calibration set to 750 samples (according to [26,38]). These samples for the calibration set were taken from the respective testing sets of the MNIST, SVHN, and CIFAR-10 datasets. The remaining samples of the testing sets were used as the in-distribution datasets for the experiment (MNIST: 9,250 samples, SVHN: 25,282 samples, CIFAR-10: 9,250 samples). Furthermore, LACA was tested in comparison to the DkNN method on a natural out-of-distribution dataset and several adversarial datasets. As the natural out-of-distribution dataset, we used the KMNIST testing set [5] for the MNIST model (10,000 samples), the CIFAR-10 testing [19] set for the SVHN model (10,000 samples), and the SVHN testing set [36] for the CIFAR-10 model (26,032 samples). To create the adversarial datasets, on the other hand, we used the FGSM attack method [9], the BIM attack

Table 2. Mean credibility scores of the in-distribution dataset (InDist), the natural out-of-distribution dataset (NatOoD) (MNIST: KMNIST, SVHN: CIFAR-10, CIFAR-10: SVHN), and the adversarial datasets (FGSM, BIM, PGD) with respect to the MNIST, SVHN, and CIFAR-10 model. For the in-distribution datasets, a higher value is better, while for the out-of-distribution datasets, a lower value is better.

Dataset	Method	InDist	NaOoD	FGSM	BIM	PGD
MNIST	DkNN	0.799	0.081	0.136	0.085	0.087
	LACA ($t = 0.01$)	0.888	0.164	0.237	0.209	0.153
	LACA ($t = 0.05$)	0.881	0.125	0.178	0.166	0.123
	LACA ($t = 0.1$)	0.880	0.124	0.173	0.166	0.121
SVHN	DkNN	0.501	0.146	0.236	0.309	0.296
	LACA ($t = 0.01$)	0.702	0.445	0.574	0.651	0.635
	LACA ($t = 0.05$)	0.634	0.233	0.390	0.500	0.474
	LACA ($t = 0.1$)	0.452	0.146	0.226	0.289	0.274
CIFAR-10	DkNN	0.522	0.229	0.168	0.179	0.221
	LACA ($t = 0.01$)	0.848	0.565	0.297	0.521	0.571
	LACA ($t = 0.05$)	0.751	0.453	0.176	0.445	0.461
	LACA ($t = 0.1$)	0.358	0.149	0.042	0.025	0.031

method [21], and the PGD attack method [32]. Each adversarial attack method was applied on the samples of the MNIST testing set, the SVHN testing set, and the CIFAR-10 testing set to create the respective adversarial datasets (using the *torchattacks* library [17]): adversarial MNIST created by FGSM ($\epsilon = 0.25$), adversarial MNIST created by BIM ($\epsilon = 0.25$, $\alpha = 0.01$, $i = 100$), adversarial MNIST created by PGD ($\epsilon = 0.2$, $\alpha = 2/255$, $i = 40$), adversarial SVHN created by FGSM ($\epsilon = 0.05$), adversarial SVHN created by BIM ($\epsilon = 0.05$, $\alpha = 0.005$, $i = 20$), adversarial SVHN created by PGD ($\epsilon = 0.04$, $\alpha = 2/255$, $i = 40$), adversarial CIFAR-10 created by FGSM ($\epsilon = 0.1$), adversarial CIFAR-10 created by BIM ($\epsilon = 0.1$, $\alpha = 0.05$, $i = 20$), and adversarial CIFAR-10 created by PGD ($\epsilon = 0.3$, $\alpha = 2/255$, $i = 40$). The accuracies achieved by the MNIST, SVHN, and CIFAR-10 model on the corresponding in-distribution and out-of-distribution datasets are shown in Table 1.

Finally, we tested the LACA method in comparison to the DkNN method on each dataset. We calculated the credibility score for every sample of a dataset using the DkNN method and took the average over all scores to obtain the mean credibility score for the dataset. Then, we repeated this calculation using the LACA method with different values for parameter t (0.01, 0.05, 0.1). For MNIST and SVHN, we computed the credibility scores of the in-distribution and the out-of-distribution datasets using the activations of all hidden layers of the model, i.e., the activations of the three convolutional layers (after ReLU) and the activations of the fully-connected layer. For CIFAR-10, on the other hand, we computed the credibility scores of the in-distribution and the out-of-distribution datasets using the activations of the first convolutional layer (after ReLU), the activations of three ResNet blocks [12] (the third, sixth, and ninth block), the activations of the global average pooling layer, and the activations of the

Table 3. Differences between the mean credibility score of the in-distribution dataset (InDist) and each out-of-distribution dataset with respect to the MNIST, SVHN, and CIFAR-10 model. We used a natural out-of-distribution dataset (NatOoD) (MNIST: KMNIST, SVHN: CIFAR-10, CIFAR-10: SVHN) and several adversarial datasets (FGSM, BIM, PGD) for each model. A higher difference value is better.

Dataset	Method	NaOoD	FGSM	BIM	PGD
MNIST	DkNN	0.718	0.663	0.714	0.712
	LACA ($t = 0.01$)	0.724	0.651	0.679	0.735
	LACA ($t = 0.05$)	**0.757**	0.703	**0.716**	0.758
	LACA ($t = 0.1$)	0.756	**0.706**	0.713	**0.759**
SVHN	DkNN	0.355	**0.265**	**0.192**	**0.205**
	LACA ($t = 0.01$)	0.257	0.128	0.051	0.067
	LACA ($t = 0.05$)	**0.401**	0.245	0.134	0.161
	LACA ($t = 0.1$)	0.307	0.227	0.163	0.178
CIFAR-10	DkNN	0.293	0.354	**0.343**	0.301
	LACA ($t = 0.01$)	0.282	0.550	0.327	0.276
	LACA ($t = 0.05$)	**0.298**	**0.575**	0.307	0.290
	LACA ($t = 0.1$)	0.209	0.315	0.333	**0.326**

Table 4. Runtimes (in seconds) of the credibility calculations of the in-distribution dataset (InDist), the natural out-of-distribution dataset (NatOoD) (MNIST: KMNIST, SVHN: CIFAR-10, CIFAR-10: SVHN), and the adversarial datasets (FGSM, BIM, PGD) with respect to the MNIST, SVHN, and CIFAR-10 model. A lower value is better.

Dataset	Method	InDist	NaOoD	FGSM	BIM	PGD
MNIST	DkNN	314.7	267.2	308.0	303.0	321.6
	LACA ($t = 0.01$)	16.5	17.9	18.7	21.4	20.6
	LACA ($t = 0.05$)	16.2	**17.2**	**17.7**	**17.8**	**17.5**
	LACA ($t = 0.1$)	**16.0**	17.5	20.0	18.6	18.5
SVHN	DkNN	1171.7	340.9	1109.3	1168.7	1197.5
	LACA ($t = 0.01$)	57.1	28.3	60.2	53.5	**46.2**
	LACA ($t = 0.05$)	**38.6**	**21.3**	43.4	42.8	50.1
	LACA ($t = 0.1$)	42.7	23.5	59.7	61.6	53.0
CIFAR-10	DkNN	684.1	1620.7	710.6	700.2	684.5
	LACA ($t = 0.01$)	47.4	137.7	52.7	53.3	55.0
	LACA ($t = 0.05$)	**42.3**	**136.0**	49.5	**49.8**	**51.1**
	LACA ($t = 0.1$)	44.4	142.7	55.6	55.5	51.4

final fully-connected layer. As a result, we obtained the mean credibility score for each in-distribution and each out-of-distribution dataset using (a) the LACA method and (b) the DkNN method. The mean credibility scores for the in-distribution datasets should be high, while the mean credibility scores for the out-of-distribution datasets should be low. The results of our experiment are shown in Table 2.

The LACA method generally obtains a similar performance compared to the DkNN method. This was also reported by Lehmann and Ebner [26]. If we set parameter t to 0.01, LACA will obtain a better result on the in-distribution datasets compared to the DkNN method. On the out-of-distribution datasets, however, the performance of the LACA method is worse than the performance of the DkNN method. Increasing parameter t improves the performance of the LACA method on the out-of-distribution datasets. Using LACA with $t = 0.1$ even obtains slightly better results compared to the DkNN method in this case. On the in-distribution datasets, on the other hand, increasing parameter t decreases the performance of LACA. Thus, it is difficult to compare the LACA method and the DkNN method as we always need to consider the performance of a method on an out-of-distribution dataset relative to its performance on the corresponding in-distribution dataset. If we obtain a low mean credibility score on the out-of-distribution dataset but also a low mean credibility score on the in-distribution dataset, for instance, we cannot distinguish between in-distribution and out-of-distribution samples (despite the desired low mean credibility score on the out-of-distribution dataset). Thus, to be able to better compare the results of the LACA and the DkNN method, we also calculated the differences between the mean credibility score of each out-of-distribution dataset and the mean credibility score of the corresponding in-distribution dataset for MNIST, SVHN, and CIFAR-10. A high difference value means that it is

Table 5. Mean credibility scores of the in-distribution dataset (InDist), the natural out-of-distribution dataset (NatOoD) (MNIST: KMNIST, SVHN: CIFAR-10, CIFAR-10: SVHN), and the adversarial datasets (FGSM, BIM, PGD) with respect to the MNIST, SVHN, and CIFAR-10 model. The credibility calculation with all layers is compared to the credibility calculation with fewer layers by unselecting layers from the front. Column *Layers* specifies the number of layers to unselect. For the in-distribution datasets, a higher value is better, while for the out-of-distribution datasets, a lower value is better.

Dataset	Method	Layers	InDist	NaOoD	FGSM	BIM	PGD
MNIST	DkNN	0	0.799	0.081	0.136	0.085	0.087
		1	0.876	0.124	0.201	0.171	0.132
	LACA	0	0.881	0.125	0.178	0.166	0.123
		1	0.963	0.396	0.449	0.425	0.288
SVHN	DkNN	0	0.501	0.146	0.236	0.309	0.296
		1	0.535	0.155	0.245	0.349	0.325
	LACA	0	0.634	0.233	0.390	0.500	0.474
		1	0.705	0.280	0.462	0.607	0.568
CIFAR-10	DkNN	0	0.522	0.229	0.168	0.179	0.221
		1	0.519	0.176	0.173	0.171	0.200
		2	0.518	0.134	0.160	0.174	0.203
	LACA	0	0.751	0.453	0.176	0.445	0.461
		1	0.839	0.480	0.205	0.481	0.498
		2	0.842	0.481	0.206	0.486	0.502

easier to distinguish between in-distribution and out-of-distribution samples. The obtained difference values are shown in Table 3. Table 3 shows that the LACA method slightly outperforms DkNN for MNIST and CIFAR-10. Furthermore, we also measured the runtimes at inference of the LACA method in comparison to the DkNN method. The results are shown in Table 4. As already reported by Lehmann and Ebner [26], LACA is significantly faster at inference than the DkNN method.

4.2 Testing with Fewer Layers

In the experiments from Lehmann and Ebner [26], the activations of the lower and the activations of the higher hidden layers of the model were used to calculate the credibility scores using LACA (described in Sect. 3) and DkNN [38]. However, the lower hidden layers, usually contain a significantly higher number of activations than the higher hidden layers. Thus, processing the lower hidden layers is more time-consuming than processing the higher hidden layers. Hence, in a second experiment, we tested whether the activations of the lower hidden layers are really necessary for the credibility calculation. We aimed to examine whether we can further improve the runtime of the LACA method by omitting some of the lower hidden layers without significantly decreasing the performance of the method. Thus, in comparison to the experiment in

Table 6. Differences between the mean credibility score of the in-distribution dataset (InDist) and each out-of-distribution dataset with respect to the MNIST, SVHN, and CIFAR-10 model. We used a natural out-of-distribution dataset (NatOoD) (MNIST: KMNIST, SVHN: CIFAR-10, CIFAR-10: SVHN) and several adversarial datasets (FGSM, BIM, PGD) for each model. The credibility calculation with all layers is compared to the credibility calculation with fewer layers by unselecting layers from the front. Column *Layers* specifies the number of layers to unselect. A higher difference value is better.

Dataset	Method	Layers	NaOoD	FGSM	BIM	PGD
MNIST	DkNN	0	0.718	0.663	0.714	0.712
		1	0.752	0.675	0.705	0.744
	LACA	0	**0.757**	**0.703**	**0.716**	**0.758**
		1	0.567	0.514	0.538	0.675
SVHN	DkNN	0	0.355	0.265	**0.192**	0.205
		1	0.381	**0.291**	0.187	**0.211**
	LACA	0	0.401	0.245	0.134	0.161
		1	**0.425**	0.243	0.098	0.136
CIFAR-10	DkNN	0	0.293	0.354	0.343	0.301
		1	0.343	0.346	0.348	0.319
		2	**0.384**	0.358	0.345	0.315
	LACA	0	0.298	0.575	0.307	0.290
		1	0.359	0.634	**0.358**	**0.341**
		2	0.361	**0.636**	0.356	0.340

Table 7. Runtimes (in seconds) of the credibility calculations of the in-distribution dataset (InDist), the natural out-of-distribution dataset (NatOoD) (MNIST: KMNIST, SVHN: CIFAR-10, CIFAR-10: SVHN), and the adversarial datasets (FGSM, BIM, PGD) with respect to the MNIST, SVHN, and CIFAR-10 model. The credibility calculation with all layers is compared to the credibility calculation with fewer layers by unselecting layers from the front. Column *Layers* specifies the number of layers to unselect. A lower runtime is better.

Dataset	Method	Layers	InDist	NaOoD	FGSM	BIM	PGD
MNIST	DkNN	0	314.7	267.2	308.0	303.0	321.6
		1	90.1	63.0	78.3	75.8	82.0
	LACA	0	16.2	17.2	17.7	17.8	17.5
		1	**12.0**	**12.3**	**12.5**	**12.9**	**13.3**
SVHN	DkNN	0	1171.7	340.9	1109.3	1168.7	1197.5
		1	348.7	116.2	319.7	344.2	334.8
	LACA	0	38.6	21.3	43.4	42.8	50.1
		1	**26.8**	**16.6**	**28.0**	**28.1**	**28.5**
CIFAR-10	DkNN	0	684.1	1620.7	710.6	700.2	684.5
		1	398.9	1107.4	465.5	439.8	445.9
		2	172.5	447.3	182.1	182.2	182.3
	LACA	0	42.3	136.0	49.5	49.8	51.1
		1	**39.0**	**100.2**	**44.8**	**43.6**	**44.6**

Sect. 4.1, we excluded some of the selected lower hidden layers for the credibility calculation and measured the impact of omitting these layers on the computed credibility scores. The results of the experiment are shown in Table 5 (mean credibility scores) and Table 6 (difference values between each out-of-distribution dataset and the corresponding in-distribution dataset). The column *Layers* of both Tables specifies whether we excluded no layers, the first layer, or the first two layers of the selected hidden layers used in the experiment in Sect. 4.1. As shown in Table 6, for the MNIST and the SVHN model we already obtained a significant performance decline when we excluded the first selected hidden layer (i.e., the first convolutional layer of the model). However, for the CIFAR-10 model, the performance slightly improved on the adversarial datasets when we excluded the first selected hidden layer (i.e., the first convolutional layer) or the first two selected hidden layers (i.e., the first convolutional layer and the first ResNet block) while achieving a faster runtime at inference (as shown in Table 7).

4.3 Testing on ImageNet Data

In the experiments from Lehmann and Ebner [26], the LACA method (described in Sect. 3) was tested in comparison to the DkNN method [38] only on simple datasets (MNIST, SVHN, CIFAR-10). Thus, in another experiment, we examined the performance of the LACA method in comparison to the DkNN method on more complex datasets: the Imagenette[1] dataset and the Imagewoof[2] dataset. Both datasets are a subset of ten classes from the ImageNet dataset [41]. Imagenette and Imagewoof do not share any classes. However, both datasets contain images of different sizes. Model training and credibility calculation, on the other hand, require a common image size. Thus, we resized all images to a common size of 128×128 beforehand.

Then, we needed to train a model for each dataset using their respective training sets (Imagenette: 9,469 training samples, Imagewoof: 9,025 training samples). For both datasets, we used a standard 18-layer residual network (ResNet) architecture [12]. To initialize the model weights, we chose a standard Kaiming Uniform [11] initialization. To extend the variety of the training images of both datasets, we added two data augmentation techniques to the training setup (random horizontal flip, random crop). We trained the Imagenette model for 40 epochs and the Imagewoof model for 60 epochs. Both models were trained using the Adam optimizer [18] and a one-cycle learning rate (LR) schedule (max LR: 0.006) [43].

We set the size of the calibration set to 750 samples. These 750 samples were taken from the respective testing sets of the Imagenette dataset and the Imagewoof dataset. The remaining samples of the testing sets were used as the in-distribution datasets for the experiment (Imagenette: 3,175 samples, Imagewoof: 3,179 samples). As in the experiments from Lehmann and Ebner [26], LACA was tested in comparison to the DkNN method on a natural out-of-distribution dataset and several adversarial datasets. As the natural out-of-distribution dataset, we simply used the Imagewoof testing set for the Imagenette model (3,929 samples), and the Imagenette testing set for the Imagewoof model (3,925 samples). To create the adversarial datasets, on the other hand, we used the FGSM attack method [9], the BIM attack method [21], and the PGD attack method [32]. Each adversarial attack method was applied on the samples of the Imagenette testing set and the Imagewoof testing set to create the respective adversarial datasets (using the *torchattacks* library [17]): adversarial Imagenette created by FGSM ($\epsilon = 0.25$), adversarial Imagenette created by BIM ($\epsilon = 0.25$, $\alpha = 0.01$, $i = 20$), adversarial Imagenette created by PGD ($\epsilon = 0.1$, $\alpha = 2/255$, $i = 40$), adversarial Imagewoof created by FGSM ($\epsilon = 0.3$), adversarial Imagewoof created by BIM ($\epsilon = 0.1$, $\alpha = 0.005$, $i = 20$), and adversarial Imagewoof created by PGD ($\epsilon = 0.1$, $\alpha = 2/255$, $i = 40$). The accuracies achieved by the Imagenette and Imagewoof model on the corresponding in-distribution and out-of-distribution datasets are shown in Table 8.

Table 8. Accuracies achieved by the Imagenette and Imagewoof model on the respective in-distribution dataset (InDist), natural out-of-distribution dataset (NatOoD) (Imagenette: Imagewoof, Imagewoof: Imagenette), and adversarial datasets (FGSM, BIM, PGD).

Model	InDist	NaOoD	FGSM	BIM	PGD
Imagenette	0.8617	0.0926	0.1231	0.0394	0.0545
Imagewoof	0.7502	0.1248	0.1217	0.0239	0.0009

---

[1] https://github.com/fastai/imagenette#imagenette-1.

[2] https://github.com/fastai/imagenette#imagewoof.

Table 9. Mean credibility scores of the in-distribution dataset (InDist), the natural out-of-distribution dataset (NatOoD) (Imagenette: Imagewoof, Imagewoof: Imagenette), and the adversarial datasets (FGSM, BIM, PGD) with respect to the Imagenette and Imagewoof model. For the in-distribution datasets, a higher value is better, while for the out-of-distribution datasets, a lower value is better.

Dataset	Method	InDist	NaOoD	FGSM	BIM	PGD
Imagenette	DkNN	0.757	0.746	0.749	0.763	0.749
	LACA ($t = 0.01$)	0.610	0.429	0.293	0.364	0.364
	LACA ($t = 0.05$)	0.476	0.295	0.218	0.204	0.193
	LACA ($t = 0.1$)	0.320	0.167	0.014	0.109	0.110
Imagewoof	DkNN	0.901	0.905	0.924	0.912	0.911
	LACA ($t = 0.01$)	0.676	0.415	0.353	0.586	0.582
	LACA ($t = 0.05$)	0.578	0.235	0.134	0.435	0.477
	LACA ($t = 0.1$)	–	–	–	–	–

Finally, we tested the LACA method in comparison to the DkNN method on each dataset. We calculated the credibility score for every sample of a dataset using the DkNN method and took the average over all scores to obtain the mean credibility score for the dataset. Then, we repeated this calculation using the LACA method with different values for parameter t (0.01, 0.05, 0.1). For Imagenette and Imagewoof, we computed the credibility scores of the in-distribution and the out-of-distribution datasets using the activations of the first convolutional layer (after ReLU), the activations of the maxpooling layer, the activations of the eight ResNet blocks [12], and the activations of the global average pooling layer. As a result, we obtained the mean credibility score for each in-distribution and each out-of-distribution dataset using (a) the LACA method and (b) the DkNN method. The mean credibility scores for the in-distribution datasets should be high, while the mean credibility scores for the out-of-distribution datasets should be low. The results of our experiment are shown in Table 9.

Table 10. Differences between the mean credibility score of the in-distribution dataset (InDist) and each out-of-distribution dataset with respect to the Imagenette and Imagewoof model. We used a natural out-of-distribution dataset (NatOoD) (Imagenette: Imagewoof, Imagewoof: Imagenette) and several adversarial datasets (FGSM, BIM, PGD) for each model. A higher difference value is better.

Dataset	Method	NaOoD	FGSM	BIM	PGD
Imagenette	DkNN	0.011	0.007	−0.006	0.008
	LACA ($t = 0.01$)	**0.181**	**0.317**	0.246	0.246
	LACA ($t = 0.05$)	**0.181**	0.258	**0.272**	**0.283**
	LACA ($t = 0.1$)	0.154	0.307	0.211	0.210
Imagewoof	DkNN	−0.004	−0.023	−0.011	−0.009
	LACA ($t = 0.01$)	0.261	0.323	0.090	0.094
	LACA ($t = 0.05$)	**0.343**	**0.444**	**0.143**	**0.101**
	LACA ($t = 0.1$)	-	-	-	-

Table 11. Runtimes (in seconds) of the credibility calculations of the in-distribution dataset (InDist), the natural out-of-distribution dataset (NatOoD) (Imagenette: Imagewoof, Imagewoof: Imagenette), and the adversarial datasets (FGSM, BIM, PGD) with respect to the Imagenette and Imagewoof model. A lower value is better.

Dataset	Method	InDist	NaOoD	FGSM	BIM	PGD
Imagenette	DkNN	2613.5	2902.8	2193.6	2336.7	2714.0
	LACA ($t = 0.01$)	196.2	281.2	207.8	**208.1**	212.5
	LACA ($t = 0.05$)	**188.9**	**277.3**	**206.7**	216.1	**194.2**
	LACA ($t = 0.1$)	193.8	278.9	211.5	212.5	207.2
Imagewoof	DkNN	1955.5	2368.2	1960.1	1989.7	1970.2
	LACA ($t = 0.01$)	**212.1**	**285.0**	219.3	212.2	199.6
	LACA ($t = 0.05$)	213.6	286.5	**200.2**	**192.1**	**197.8**
	LACA ($t = 0.1$)	–	–	–	–	–

As in Sect. 4.1, to be able to better compare the results of the LACA and the DkNN method, we also calculated the differences between the mean credibility score of each out-of-distribution dataset and the mean credibility score of the corresponding in-distribution dataset for Imagenette and Imagewoof. A high difference value means that it is easier to distinguish between in-distribution and out-of-distribution samples. The obtained difference values are shown in Table 10. Table 10 shows that the LACA method produces meaningful results. However, LACA performed slightly worse on Imagenette and Imagewoof compared to the simple datasets (Sect. 4.1). Furthermore, a value of 0.1 for parameter t was already too high for Imagewoof. As a result, the mean credibility score of the Imagewoof in-distribution dataset was 0 in this case. The DkNN method, on the other hand, did not produce any meaningful credibility scores. The scores for the in-distribution datasets and the out-of-distribution datasets were similar. Thus, in contrast to LACA, the DkNN method cannot be used for distinguishing between in-distribution samples and out-of-distribution samples from Imagenette and Imagewoof. Additionally, the LACA method was also significantly faster than the DkNN method (as shown in Table 11).

5   Conclusion

In Sect. 1, we stated the following research question: Can we further decrease the runtime of the LACA method by omitting some of the hidden layers for the credibility calculation without significantly lowering its performance? Our experiment in Sect. 4.2 showed that omitting layers from the credibility calculation results in a significant performance loss in most cases. However, for the adversarial datasets with respect to the CIFAR-10 model, we observed a slightly improved performance. When omitting the first hidden layer from the credibility calculation, for instance, the mean credibility scores of the adversarial datasets showed a slight increase, but the mean credibility score of the in-distribution dataset showed an even stronger increase at the same time.

Thus, the ability to distinguish between in-distribution samples and adversarial samples improved in this case. We assume that this result is caused by the behavior of adversarial samples. We observed visually that at the lower hidden layers an adversarial sample often still appears to be close to in-distribution samples of the same class as the original sample from which the adversarial sample was created. However, at the middle hidden layers, the adversarial sample is suddenly close to in-distribution samples of a different class. We should probably not omit the layer where this occurs when calculating the credibility. However, the CIFAR-10 model used in Sect. 4 contains a high number of layers. Thus, we assume the first few lower hidden layers of the model are not that important for the credibility calculation. Furthermore, in Sect. 1, we also stated a second research question: Does the LACA method also work for more complex datasets in a reasonable time compared to the DkNN method? Our experiment in Sect. 4.3 showed that the LACA method does work on more complex datasets such as the Imagenette dataset and the Imagewoof dataset (subsets of the ImageNet dataset [41]). LACA was significantly faster at inference than the DkNN method. Furthermore, we obtained meaningful credibility scores using LACA. The DkNN method, on the other hand, did not compute any meaningful credibility scores for the Imagenette dataset and the Imagewoof dataset. Thus, the LACA method outperforms the DkNN method in this case. Nevertheless, the performance of the LACA method on these more complex datasets was not as high as its performance for the simple datasets (Sect. 4.1). Thus, in future work, the performance of LACA on complex datasets still needs to be improved.

References

1. Biggio, B., et al.: Evasion attacks against machine learning at test time. In: Blockeel, H., Kersting, K., Nijssen, S., Železný, F. (eds.) ECML PKDD 2013. LNCS (LNAI), vol. 8190, pp. 387–402. Springer, Heidelberg (2013). https://doi.org/10.1007/978-3-642-40994-3_25
2. Carrara, F., Falchi, F., Caldelli, R., Amato, G., Becarelli, R.: Adversarial image detection in deep neural networks. Multimedia Tools Appl. **78**(3), 2815–2835 (2019)
3. Chen, B., et al.: Detecting backdoor attacks on deep neural networks by activation clustering. In: Espinoza, H., Ó hÉigeartaigh, S., Huang, X., Hernández-Orallo, J., Castillo-Effen, M. (eds.) Workshop on SafeAI@AAAI. CEUR Workshop, vol. 2301. ceur-ws.org, Honolulu, HI, USA (2019)
4. Chen, T., Navratil, J., Iyengar, V., Shanmugam, K.: Confidence scoring using whitebox meta-models with linear classifier probes. In: Chaudhuri, K., Sugiyama, M. (eds.) AISTATS, vol. 89, pp. 1467–1475. PMLR, Naha, Japan (2019)
5. Clanuwat, T., Bober-Irizar, M., Kitamoto, A., Lamb, A., Yamamoto, K., Ha, D.: Deep learning for classical Japanese literature. ArXiv arXiv:1812.01718 (2018)
6. Cohen, G., Sapiro, G., Giryes, R.: Detecting adversarial samples using influence functions and nearest neighbors. In: CVPR. pp. 14441–14450. IEEE, Seattle, WA, USA (2020)
7. Gal, Y.: Uncertainty in Deep Learning. Ph.D. thesis, Univ of Cambridge (2016)
8. Gal, Y., Ghahramani, Z.: Dropout as a bayesian approximation: representing model uncertainty in deep learning. In: Balcan, M., Weinberger, K. (eds.) ICML, vol. 48, pp. 1050–1059. PMLR, New York, NY, USA (2016)
9. Goodfellow, I., Shlens, J., Szegedy, C.: Explaining and harnessing adversarial examples. In: Bengio, Y., LeCun, Y. (eds.) ICLR. San Diego, CA, USA (2015)

10. Grosse, K., Manoharan, P., Papernot, N., Backes, M., McDaniel, P.: On the (statistical) detection of adversarial examples. ArXiv arXiv:1702.06280 (2017)
11. He, K., Zhang, X., Ren, S., Sun, J.: Delving deep into rectifiers: surpassing human-level performance on imagenet classification. In: ICCV. pp. 1026–1034. IEEE, Santiago, Chile (2015)
12. He, K., Zhang, X., Ren, S., Sun, J.: Deep residual learning for image recognition. In: CVPR, pp. 770–778. IEEE, Las Vegas, NV, USA (2016)
13. Hendrycks, D., Gimpel, K.: A baseline for detecting misclassified and out-of-distribution examples in neural networks. In: ICLR, Toulon, France (2017)
14. Hendrycks, D., Mazeika, M., Kadavath, S., Song, D.: Using self-supervised learning can improve model robustness and uncertainty. In: Wallach, H., Larochelle, H., Beygelzimer, A., d'Alché-Buc, F., Fox, E., Garnett, R. (eds.) NeurIPS, vol. 32, pp. 15637–15648. CAI, Vancouver, CA (2019)
15. Hendrycks, D., Zhao, K., Basart, S., Steinhardt, J., Song, D.: Natural adversarial examples. ArXiv arXiv:1907.07174 (2020)
16. Huang, H., Li, Z., Wang, L., Chen, S., Dong, B., Zhou, X.: Feature space singularity for out-of-distribution detection. In: Espinoza, H., et al., (eds.) Workshop on SafeAI@AAAI. CEUR Workshop, vol. 2808. ceur-ws.org (2021)
17. Kim, H.: Torchattacks: A pytorch repository for adversarial attacks. ArXiv arXiv:2010.01950 (2020)
18. Kingma, D.P., Ba, J.: Adam: a method for stochastic optimization. In: Bengio, Y., LeCun, Y. (eds.) ICLR. San Diego, CA, USA (2015)
19. Krizhevsky, A.: Learning multiple layers of features from tiny images. University of Toronto, Technical Report (2009)
20. Krizhevsky, A., Sutskever, I., Hinton, G.E.: Imagenet classification with deep convolutional neural networks. In: Pereira, F., Burges, C.J.C., Bottou, L., Weinberger, K.Q. (eds.) NIPS, vol. 25, pp. 1097–1105. CAI, Lake Tahoe, NV, USA (2012)
21. Kurakin, A., Goodfellow, I.J., Bengio, S.: Adversarial examples in the physical world. In: ICLR. Toulon, France (2017)
22. LeCun, Y., Cortes, C., Burges, C.: Mnist handwritten digit database. ATT Labs. https://yann.lecun.com/exdb/mnist 2 (2010)
23. Lee, K., Lee, H., Lee, K., Shin, J.: Training confidence-calibrated classifiers for detecting out-of-distribution samples. In: ICLR. Vancouver, CA (2018)
24. Lee, K., Lee, K., Lee, H., Shin, J.: A simple unified framework for detecting out-of-distribution samples and adversarial attacks. In: Bengio, S., Wallach, H., Larochelle, H., Grauman, K., Cesa-Bianchi, N., Garnett, R. (eds.) NeurIPS, vol. 31, pp. 7167–7177. CAI, Montreal, CA (2018)
25. Lehmann, D., Ebner, M.: Layer-wise activation cluster analysis of CNNs to detect out-of-distribution samples. In: Farkaš, I., Masulli, P., Otte, S., Wermter, S. (eds.) ICANN 2021. LNCS, vol. 12894, pp. 214–226. Springer, Cham (2021). https://doi.org/10.1007/978-3-030-86380-7_18
26. Lehmann, D., Ebner, M.: Calculating the credibility of test samples at inference by a layer-wise activation cluster analysis of convolutional neural networks. In: Proceedings of the 3rd International Conference on Deep Learning Theory and Applications DeLTA 2022, pp. 34–43. INSTICC, SciTePress, Lisbon, Portugal (2022)
27. Lehmann, D., Ebner, M.: Subclass-based under sampling for class-imbalanced image classification. In: Proceedings of the 17th International Joint Conference on Computer Vision. Imaging and Computer Graphics Theory and Applications - Volume 5: VISAPP, pp. 493–500. SciTePress, INSTICC (2022)
28. Li, X., Li, F.: Adversarial examples detection in deep networks with convolutional filter statistics. In: ICCV, pp. 5775–5783. IEEE, Venice, Italy (2017)

29. Lin, Z., Roy, S.D., Li, Y.: Mood: multi-level out-of-distribution detection. In: CVPR, pp. 15308–15318. IEEE (2021)
30. Ma, X., et al.: Characterizing adversarial subspaces using local intrinsic dimensionality. In: ICLR, Vancouver, CA (2018)
31. MacQueen, J.B.: Some methods for classification and analysis of multivariate observations. In: Cam, L.M.L., Neyman, J. (eds.) Berkeley Symposium on Mathematical Statistics and Probability, vol. 1, pp. 281–297. University of California Press (1967)
32. Madry, A., Makelov, A., Schmidt, L., Tsipras, D., Vladu, A.: Towards deep learning models resistant to adversarial attacks. In: ICLR. Vancouver, CA (2018)
33. McInnes, L., Healy, J., Melville, J.: UMAP: Uniform manifold approximation and projection for dimension reduction. ArXiv arXiv:1802.03426 (2018)
34. Meng, D., Chen, H.: Magnet: a two-pronged defense against adversarial examples. In: SIGSAC, pp. 135–147. ACM, Dallas, TX, USA (2017)
35. Metzen, J.H., Genewein, T., Fischer, V., Bischoff, B.: On detecting adversarial perturbations. In: ICLR, Toulon, France (2017)
36. Netzer, Y., Wang, T., Coates, A., Bissacco, A., Wu, B., Ng, A.Y.: Reading digits in natural images with unsupervised feature learning. In: NIPS Workshop on Deep Learning and Unsupervised Feature Learning (2011)
37. Nguyen, A., Yosinski, J., Clune, J.: Multifaceted feature visualization: uncovering the different types of features learned by each neuron in deep neural networks. Visualization for Deep Learning workshop. In: International Conference in Machine Learning (2016). arXiv preprint arXiv:1602.03616
38. Papernot, N., McDaniel, P.: Deep k-nearest neighbors: towards confident, interpretable and robust deep learning. ArXiv arXiv:1803.04765 (2018)
39. Pearson, K.: LIII. On lines and planes of closest fit to systems of points in space. London, Edinburgh Dublin Philos. Mag. J. Sci. 2(11), 559–572 (1901)
40. Rousseeuw, P.J.: Silhouettes: a graphical aid to the interpretation and validation of cluster analysis. J. Comput. Appl. Math. 20(1), 53–65 (1987)
41. Russakovsky, O., et al.: Imagenet large scale visual recognition challenge. IJCV 115(3), 211–252 (2015)
42. Sastry, C.S., Oore, S.: Detecting out-of-distribution examples with gram matrices. In: ICML, vol. 119, pp. 8491–8501. PMLR (2020)
43. Smith, L.N.: Cyclical learning rates for training neural networks. In: WACV, pp. 464–472. IEEE (2017)
44. Szegedy, C., et al.: Intriguing properties of neural networks. In: Bengio, Y., LeCun, Y. (eds.) ICLR. Banff, CA (2014)
45. Zeiler, M.D., Fergus, R.: Visualizing and understanding convolutional networks. In: Fleet, D., Pajdla, T., Schiele, B., Tuytelaars, T. (eds.) ECCV 2014. LNCS, vol. 8689, pp. 818–833. Springer, Cham (2014). https://doi.org/10.1007/978-3-319-10590-1_53
46. Zhang, H., Dauphin, Y.N., Ma, T.: Fixup initialization: residual learning without normalization. ArXiv arXiv:1901.09321 (2019)

Traffic Sign Repositories: Bridging the Gap Between Real and Synthetic Data

Diogo Lopes da Silva[1] and António Ramires Fernandes[2]([⊠]) [iD]

[1] Universidade do Minho, Braga, Portugal
[2] Algoritmi Centre/Department of Informatics/LASI, Universidade do Minho, Braga, Portugal
arf@di.uminho.pt

Abstract. Creating a traffic sign dataset with real data can be a daunting task. We discuss the issues and challenges of real traffic sign datasets, and evaluate these issues from the perspective of creating a synthetic traffic sign dataset. A proposal is presented, and thoroughly tested, for a pipeline to generate synthetic samples for traffic sign repositories. This pipeline introduces Perlin noise and explores a new type of noise: Confetti noise. Our pipeline is capable of producing synthetic data which can be used to train models producing state of the art results in three public datasets, clearly surpassing all previous results with synthetic data. When merged or ensemble with real data our results surpass previous state of the art reports in three datasets: GTSRB, BTSC, and rMASTIF. Furthermore, we show that while models trained with real data datasets perform better in the respective dataset, the same is not true in general when considering other similar test sets, where models trained with our synthetic datasets surpassed models trained with real data. These results hint that synthetic datasets may provide better generalization than real data, when the testing data is outside of the distribution of the real data.

Keywords: Synthetic data · Traffic sign classification · Convolutional Neural Networks

1 Introduction

Creating a representative recognition traffic sign repository is an appreciable challenge, requiring gathering and labelling samples for a large number of classes, under a variety of lighting and deterioration conditions. To aggravate things further, we must also consider the potential intraclass variability due to multiple versions of the same sign. Furthermore, pictograms and fonts vary from country to country, implying that the usage of a dataset across different countries will come with an associated performance cost. Building a representative multinational dataset will imply that the number of classes explodes, which in turn implies a much larger dataset, and potentially a larger supporting model.

Inspecting publicly available European traffic signs we find a reduced number of classes, and a significant number of classes with a low number of samples. Given a traffic sign dataset, some works address the accuracy problem from the model architecture perspective. Complex architectures, namely Spatial Transformer Networks (STN), Inception modules, and Generative Adversarial Networks (GAN), have been able to achieve

A. Fred et al. (Eds.): DeLTA 2022, CCIS 1858, pp. 56–77, 2023.
https://doi.org/10.1007/978-3-031-37317-6_4

considerable accuracies. It is important to note, however, that the test sets for each particular traffic sign repository are commonly similar in terms of lighting and atmospheric conditions to respective the training sets. Hence, the accuracies reported in these works do not necessarily generalize to new samples captured under different conditions.

Another approach is to focus on the data. The usage of synthetic data for traffic sign repositories has been addressed multiple times in the literature [5,6,9–11,17,19]. Synthetic datasets eliminate several of the issues mentioned above. Nevertheless, the issue of getting a representative dataset remains relevant. While in synthetic datasets, the gathering and labelling processes are no longer required, simulating lighting and deterioration conditions remains a challenge.

The only required input for a synthetic dataset in [1,10,19] is a set of templates for each class. A set of geometric and colour operations are then applied to provide sample diversity, in an attempt to achieve a truly representative dataset. Other approaches require the existence of real data, and mostly use GANs in an attempt to generate synthetic samples that belong to the same distribution as real data [5,9,11,17].

In [10] a pipeline for the generation of synthetic data for traffic sign repositories was proposed. The results obtained provide some hope that the gap between real and synthetic data can be closed for traffic sing repositories. The proposal does not require real data, but it shows how, given the availability of such data, performance can be increased for some datasets.

The present work is an extended version of [10], where we expand the discussion relating to real vs. synthetic data, detailing the issues that each approach faces. We also provide extended testing, adding detailed testing on the new operators proposed, adding new datasets, and extended our analysis of the results, allowing us to consolidate our conclusions.

2 Synthetic vs. Real Traffic Sign Datasets

Gathering enough data for a real traffic sign dataset is both a time and resource consuming task, with well over 100 different traffic signs classes requiring collecting a significant number of samples per class.

The following issues concerning real traffic sign datasets were identified during this work:

– Scarcity: some signs are rare in particular countries. For instance, the diamond shape yield sign is uncommon in Portugal. Collecting a significant number of these signs might be unfeasible.
– Placement: some signs can only be found at certain regions, for instance, the warning sign for snow. This implies that travelling is required to those regions to gather samples.
– Lighting: lighting varies along the day, and seasonally as well. Even if we dismiss the seasonal variation, the intraday variation (including cloudy skies) can affect the accuracy of a model.
– Weather exposure, graffiti and stickers: these are all elements that can degrade significantly the ability to recognise a sign. See Fig. 1 for some examples from Belgium.

- Adverse atmospheric conditions: the presence of rain, snow and fog should be taken into account when gathering samples to provide high accuracy in what are probably the most demanding situations for the driver.
- Intraclass variation: Traffic signs within the same class can have different pictograms, or use different fonts. This is mostly due to the introduction of new versions over time for some classes. Furthermore, manufacturing issues can also cause differences in the pictograms. See Fig. 2 for some examples from German traffic signs.
- Maintenance: When new traffic signs are introduced, it will take some time until sufficient samples are gathered to retrain the model.
- Camera equipment: different sensors will produce samples which will differ in contrast, brightness and hue.

Fig. 1. Sample of signs from Belgium that show some degradation such as weathering, graffiti, and stickers.

Fig. 2. Sample of intraclass variations per country in GTSRB. Letf and middle: variation due to new templates being introduced; Right: variation due to manufacturing.

Another issue is the absence of an international observed standard for traffic sign pictograms. Traffic signs can vary significantly from country to country as can be seen when inspecting the ETSD, which contains samples from six European countries. Some examples are provided in Fig. 3. This implies that simply using a traffic sign dataset across countries is not an option, although some classes can be reused.

Fig. 3. Sample of intraclass variations that can be found across different countries.

While all these issues do not impair our ability to correctly interpret a traffic sign, they can have a severe effect on a deep learning model accuracy.

Regarding synthetic traffic signs, and considering building a synthetic dataset for a country, scarcity is not an issue, as there is no inherent limitation on the number of samples to be synthesised. Similarly, there is no issue with particular signs only being found in certain regions. Intraclass variation is also easily tackled as multiple templates can be used per class. Maintenance is required each time a new template is added to the dataset, but for synthetic signs only retraining the model is needed, and this can be done as early as when the new sign is designed.

The real issues when building synthetic traffic sign datasets are related to lighting, degradation, atmospheric conditions, and differences in sensors.

Being able to simulate these conditions will allow for a well designed dataset. This is not a simple task, however, once achieved, building traffic sign datasets for any country is only a matter of gathering the respective templates and train a model.

3 Related Work

This section focuses on the European traffic sign repositories used in this work, and on research relating to synthetic traffic sign datasets.

3.1 Traffic Sign Datasets

Only for a few countries have traffic sign samples been collected, labelled, and released as a public dataset. The volume and quality of samples varies greatly across countries.

This works focuses mainly on three European datasets: GTSRB[1] [18] (Germany); BTSC[2] [20] (Belgium); and rMASTIF[3] [23] (Croatia).

Table 1 presents some statistics for these datasets.

Table 1. Statistics for the German, Belgian, and Croatian datasets.

	GTSRB	BTSC	rMASTIF
class #	43	62	31
train #	39209	4575	4044
test #	12630	2520	1784
min res	25x25	22x21	17x16
max res	232x266	674x527	185x159

In [3], Serna and Ruichek propose the European Traffic Sign Dataset (ETSD). This dataset merges existing datasets from six European Countries: Belgium, Croatia, France, Germany, Netherlands, and Sweden. For Croatia and Germany, besides using the previously identified datasets the authors also took advantage of the respective detection datasets to gather more samples. Regarding France, Netherlands, and Sweden, the datasets used were:

- Stereopolis dataset (France) [14];
- UTBM (France [3]
- RUG Traffic sign image database (Netherlands) [2]
- STS (Sweden) [8]

---

[1] http://benchmark.ini.rub.de/.

[2] https://btsd.ethz.ch/shareddata/.

[3] http://www.zemris.fer.hr/~kalfa/Datasets/rMASTIF/.

The final ETSD dataset has 164 classes, with a training/test split of 60546/21930 samples. Data from this dataset is used for cross-testing, see Sect. 5.3.

Other datasets are available, for instance the Italian or DITS dataset [21], and the Tsinghua-Tencent 100K benchmark [22].

3.2 Synthetic Traffic Signs

Stergiou et al. [19] proposed the generation of synthetic training dataset based on traffic sign templates. Templates for 50 classes of British traffic signs were gathered, and composed with background images of British roads, both from rural and urban areas.

Processing the templates consisted in both colour and geometric processing. Regarding colour, the goal was to simulate different lighting conditions, in order to approximate real life scenarios, with the final dataset containing 4 brightness variations. Considering the geometric transformations, 20 distinct affine transformations for shearing were applied, alongside rotations, scaling, and translations.

This dataset was evaluated with a CNN model with 6 convolutional layers achieving a peak accuracy of 92.20%. However, since the test dataset has not been provided, no comparisons with other works can be made.

Luo et al. [11] approached the synthetisation of the dataset based on Generative Adversarial Networks (GAN). The authors also implemented a conventional pipeline to generate synthetic samples using both colour and geometric transformations and claim to achieve more realistic imagery with the GAN approach.

The main purpose in using GANs is that the GAN itself will learn the generation parameters from real data, as opposed to the conventional pipeline where the parameters are manually tuned. On the down side, this approach requires existing real data to train the GAN, and unless the real data is truly representative, the synthetic data will inherit biases from the real data.

As input, the algorithm receives a sign template, an affine transformation, and a background. The GAN is responsible for the synthesis of the merging background and the traffic sign template. The geometric transformations are applied independently as in Stergiou et al. [19].

An accuracy of 97.24% was achieved on a subset of the GTSRB test set, not including the yield diamond shade sign. For comparison purposes Luo et al. also presents the model accuracy when training with only real data: 99.21%. A dataset consisting of merged real and synthetic data was also tested achieving an accuracy of 99.41%, using 50% of the real training data.

Extending the proposal of Luo et al. [11], Spata et al. [17] propose to generate the background itself with a GAN. The geometric transformations are still applied independently. As the authors state, "the CycleGAN is designed primarily for stylistic and textural translations and therefore cannot effectively contribute such information itself". The reported accuracy result with synthetic datasets is 95.15%.

Horn and Houben [5] further explore the generation of synthetic data with Cycle-GANs, however, results are only provided for selected classes.

Araar et al. in [1] propose a conventional pipeline with geometric transformations and image processing techniques, as in Stergiou et al. [19]. With a DenseNet architecture, an accuracy of 97.83% in GTSRB is reported using only synthetic data.

Liu et al. [9] explore the generation of synthetic data using a DCGAN trained on real data. Their work shows that it is possible to create images with a high degree of similarity based on the SSIM metric.

Horn et al. [6] propose assessing the quality of generated synthetic samples using four different measures. The main purpose is to evaluate if a synthetic image is significantly different from the distribution of real images.

Luo and Wang [12] propose a pipeline to label real images. A synthetic dataset is produced using conventional geometric and colour operations on templates, which are then merged to real backgrounds. This dataset is then used to train a model that will be used in real unlabelled data in order to provide the labels. Their results are not directly comparable to previous works since the authors are more focused on recall and not on the performance of the synthetic dataset per se. This process is repeated and a recall of 98.6% is achieved with all images being correctly classified.

A relevant note is that all works strived to generate synthetic samples as close to real samples as possible.

4   Synthetic Traffic Signs Generation Algorithm

As in previous works, a synthetic sample is a composition of a background image with a foreground template that undergoes a set of operations.

The traffic sign synthesising algorithm is a pipeline of geometric transformations, colour transformations, and image disturbances in the form of noise and blur. To define the set of operations for our pipeline we examined real traffic sign datasets to identify the relevant operations to include.

To define the set of templates used we inspected only the training set of each dataset. A sample of the gathered templates for GTSRB is shown in Fig. 4. Some classes have multiple templates due to the presence of older versions of a sign, or even manufacturing differences (an example can be seen in the templates for 120 km/h speed limit). This is common, but not exclusive to speed limit classes. Some templates are rotated to accommodate for the real sign placement, for instance the roundabout sign.

Fig. 4. Sample templates for the GTSRB.

An interesting issue arises when traffic signs share the same pictogram, yet belonging to different classes. And example can be found in the rMASTIFF dataset. The main difference between these signs is the shape and colour of the outer area of the sign. This naturally results in a number of inter-class misclassifications. To deal with this issue multiple templates were used for each class varying the hue and luminance channels, see Fig. 5. This approach successfully reduces the number of misclassified samples from both classes.

In the remainder of this section we first discuss the background options, the usage of information from real data distribution, new operators to synthesise samples, concluding with the full pipeline presentation.

Class 10   Class 28

Fig. 5. Templates for classes with same central pictogram (rMASTIFF). Source [10].

4.1 Background

As discussed in Sect. 3.2, the usage of real scenario backgrounds in synthetic samples is the common approach. In our work, the real imagery backgrounds come from signless images from Google Street View.

As depicted in Fig. 6, for our synthetic data generation we further tested an alternative background approach: random solid colour per sample.

Fig. 6. Real vs. solid colour backgrounds. Source [10].

While real backgrounds provide more realistic imagery they may introduce a bias in the training set. Different regions share different architectural trends, and even rural areas can be very diverse. Thus, it can be challenging to find a suitable set of backgrounds covering a significant number of scenarios. Furthermore, adding weather conditions, and lighting variations due to time of day or even seasons, only aggravates this quest for having representative backgrounds.

Random solid colour backgrounds, on the other hand, avoid all the previously discussed issues, and "force" the network to focus on the traffic sign since there are no features outside of the traffic sign. This approach has been tested previously in [1], but Araar et al. discarded this option due to poor results.

In our pipeline, we explore both real ans solid colour backgrounds.

4.2 Brightness Distribution

Real image data can be modelled by statistical data distributions for some of its parameters. Brightness is the example explored in this work. The availability of real data allows to compute brightness for synthetic samples based on the real data brightness distribution.

To find a distribution that fits the real dataset brightness distribution of the real datasets, the Kolmogorov-Smirnov test (K-S test) was performed. Running the K-S test on all available samples from the three datasets, we found that the Johnson distribution with bounded values was able to closely fit the real sample data. The histogram plot of the distribution for the three main datasets explored in our work is depicted in Fig. 7. Table 2 presents the parameters of the distributions.

Fig. 7. Brightness frequency distribution for BTSC, rMASTIF, and GTSRB. The curve represents the Johnson fitted distribution. The horizontal axis represents the sample average brightness.

Table 2. Fitted Johnson brightness distribution parameters for the German, Belgian, and Croatian datasets. Source [10].

dataset	Parameter			
	γ	δ	ξ	λ
GTSRB	0.747	0.907	7.099	259.904
BTSC	0.727	1.694	2.893	298.639
rMASTIF	0.664	1.194	20.527	248.357

To adjust the brightness of a synthetic sample, a brightness value was sampled from the Johnson distribution, and the average brightness of the template is adjusted to match the sampled brightness. Examples of the end result can be seen in Fig. 8.

Fig. 8. Brightness variation in synthetic samples of class 7 for the GTSRB dataset. Source [10].

While this approach may provide trained models with higher accuracy in samples with the same brightness distribution, the usage of a brightness distribution also contaminates the synthetic dataset with the biases present in the training set. As can be seen in Fig. 7 the distribution for the three datasets varies significantly. Furthermore, this approach can only be used if real data is available. Note, however, that to compute a brightness distribution an unlabelled set of real traffic signs is sufficient.

To create a synthetic dataset from scratch we propose the usage of exponential Eq. 1, where the desired brightness is computed considering a uniform random variable u in $[0, 1]$, and a variable $bias$ that determines the minimum brightness. Brightness B can be defined in the range $[bias, 255]$ as:

$$B = bias + u^{\gamma} \times (255 - bias) \tag{1}$$

In our tests we set $bias = 10$, and $\gamma = 2$.

For both approaches the process to adjust the template brightness is identical, being performed in HSV colour space. The first step is to compute the average V component in HSV representation. A ratio between the desired brightness and the mean V value is computed, and multiplied by V for every pixel.

4.3 Confetti Noise

A significant portion of the smaller samples from the real datasets have abrupt pictogram colour variations. Some examples are presented in Fig. 9.

Fig. 9. Examples of noisy traffic sign samples of classes 0, 1, 2, and 3 from the GTSRB dataset, respectively. Source [10].

Our approach to simulate this phenomena is based on impulsive noise. This noise, which we named Confetti Noise, modifies the value of pixels in a random fashion, being applied only to the smaller samples.

Confetti noise is only applied to smaller samples and has 3 parameters. The kernel size ratio (3% of the original template dimension), the probability of updating the window under the kernel (set at 3%), and the stride (set at 1.5% of the template dimensions).

Finally, the template is resized to the desired resolution. The effect on a template and a comparison with an actual traffic sign is depicted in Fig. 10.

Fig. 10. Confetti noise. From the left: original template, resized sample, resized template after applying confetti noise, and sample from GTSRB. Source [10].

4.4 Perlin Noise

Perlin Noise [15] has been used in Computer Graphics as a technique to produce natural appearing textures. In the context of creating a synthetic dataset, Perlin Noise can be used to simulate the heterogeneity found in real traffic signs due to uneven light exposure, colour fade, or deterioration due to exposure. Hence, Perlin noise was added to our pipeline.

The process of applying Perlin noise to the templates consists first in random cropping of a large noise texture, and alpha blending the crop with the template, see Eq. 2. The Perlin noise parameters are as follows: 6 octaves, a persistence of 0.5, and a lacunarity of 2.0. Figure 11 shows examples of Perlin noise applied to different templates.

$$final = (1 - \alpha) \times template + \alpha \times noise \tag{2}$$

Fig. 11. Perlin noise sample (left) applied to classes 1, 36, and 41 from the GTSRB dataset with $\alpha = 0.4$. Source [10].

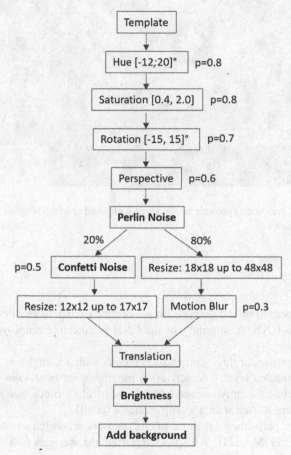

Fig. 12. Synthetic template transformation pipeline.

4.5 Synthetic Generation Pipeline

Most of the operations we use to generate synthetic samples are similar to those in [19] and [1]. The geometric operations used were resizing, translation, rotation, and perspective transforms. Regarding colour we also perform hue and saturation jitter.

The full pipeline[4] is depicted in Fig. 12. On the side of each box the probability of applying the respective transformation to a sample is displayed. Branching occurs to

---

[4] Source code for the generation of synthetic datasets available at https://github.com/Nau3D/bridging-the-gap-between-real-and-synthetic-traffic-sign-datasets.

provide a different treatment to small and large samples. Depicted in bold are the items that represent the novelty and were discussed above.

Figure 13 presents samples of synthesised samples considering solid colour backgrounds. As opposed to previous works our samples are clearly not photo-realistic.

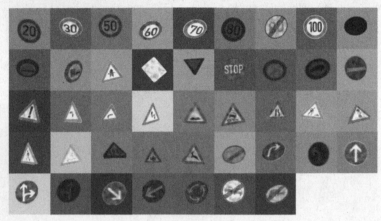

Fig. 13. Samples of generated synthetic traffic signs with solid colour backgrounds for each class of the GTSRB dataset.

5   Evaluation

As this work is focused on the datasets an not on fine tuning an architecture, we opted for a plain vanilla CNN. A summary of the CNN architecture employed in this work can be seen in Table 3.

This model consists of three convolution blocks with a kernel size of 5×5 pixels and one fully connected layer. The activation function used in all convolutional layers is LeakyReLU, while the fully connected layer uses ReLU. Batch size is set to 64, and the Adam optimizer is used with a learning rate of 0.0001.

Most tests are performed on three of the datasets presented in Sect. 3.1, namely GTSRB, BTSC, and rMASTIF. All values reported are averages of 5 runs, each completing 40 epochs.

On real datasets data augmentation has been performed prior to training to achieve a more balanced dataset, with each class ending with at least 2000 samples. To take advantage of the available data we first performed horizontal flipping where applicable. This is particularly useful when the flipped image ends up in another class, as is the case with turning signs. After this step, and for those classes still having less than 2000 samples, we performed common geometric operations, namely, translations (up to four pixels in each direction), and rotations ($-10°$ to $10°$).

Dynamic data augmentation consisting of geometric and colour operations is applied to the dataset during training. The operations involved are: rotations with a maximum of $5°$ in each direction; shear in the range of $[-2, 2]$ pixels followed by rotations; translations in a range of $[-0.1, 0.1]$ percent, also followed by rotations; and

Table 3. Neural network model with a total of approximately 2.7 million trainable parameters. The number of outputs is set according to the number of classes of the dataset. Source [10].

Layer type	Filters	Size
Input		32×32
Convolution + LeakyReLU	100	5×5
Batch Norm		
Dropout ($p = 0.05$)		
Convolution + LeakyReLU	150	5×5
Max Pooling		2×2
Batch Norm		
Dropout ($p = 0.05$)		
Convolution + LeakyReLU	250	5×5
Max Pooling		2×2
Batch Norm		
Dropout ($p = 0.05$)		
Fully Connected + ReLU		350
Fully Connected + ReLU		# classes (c)

centre cropping of 28×28 pixels. The colour transformation consists in jittering of the brightness, saturation, contrast, each multiplied by a random value in the range $[0, 3]$, and hue jittered in the range $[-0.4, 0.4]$. As mentioned in Sect. 4, when classes have similar or even identical pictograms, and only the outer colour differs, it is preferable to have multiple templates for those classes and a smaller range for hue jittering. For the Croatian rMASTIF dataset the range $[-0.2, 0.2]$ produces better results.

To increase the diversity, these transformations are applied independently, i.e., for each sample in the original dataset, eight more samples are produced in the data augmentation process.

Accuracy results for real data are reported in Table 4, including state-of-the-art results, to put in context the results obtained with synthetic data presented in the following subsections. Our results are an average of 5 runs per dataset.

Note that we are using a smaller input when compared to reference works. This results not only in a smaller memory footprint, but also a faster evaluation due to the lower number of convolution operations required.

5.1 Solo Synthetic Dataset Evaluation

Synthetic datasets have 2000 samples per class, and the dynamic data augmentation procedure is as described for real data datasets. According to Sect. 4.2 and 4.1 we have two variations for both brightness and background.

We prepared synthetic datasets to contemplate these variations resulting in 4 distinct synthetic dataset types. Brightness can be set according to the exponential equation (Eq. 1), or sampling from the respective Johnson distribution. Backgrounds can be crops of signless real images, or just a solid colour square.

Table 4. Accuracy results when training with real data. Number of parameters is 10^6.

Dataset	Model	Input	Params	Acc (%)
GTSRB	Ours	32×32	2.7	99.64 ± 0.02
	Mahmoud and Guo. [13]	64×64	–	99.80
	Haloi [4]	128×128	10.5	**99.81**
BTSC	Saha et al. [16]	56×56	6.3	99.17
	Ours	32×32	2.7	99.30 ± 0.03
	Mahmoud and Guo. [13]	64×64	–	**99.72**
rMASTIF	Jurišić et al. [7]	48×48	6.3	99.53
	Ours	32×32	2.7	**99.71** ± 0.05

To identify the dataset type, we use **R** for real data datasets. Synthetic datasets are identified by a three letter abbreviation, always starting with **S** for synthetic. The second letter relates to brightness option and the third to the background used. The synthetic dataset types are referred to as:

- **SES** - **E**xponential brightness and **S**olid bgs;
- **SJS** - **J**ohnson brightness dist. and **S**olid bgs;
- **SER** **E**xponential brightness and **R**eal bgs;
- **SJR** - **J**ohnson brightness dist. and **R**eal bgs;

To provide more meaningful results, for each type we created five datasets varying a random seed.

Perlin noise was the first feature to be tested. Regarding our pipeline we turned off Confetti noise, and experiment with different intensities of Perlin noise, i.e., with different α values in Eq. 2. The **SES** synthetic datasets were used for this test. Results reported in Table 5, clearly show the advantage of using Perlin noise, with a very significant increase in accuracy when comparing $\alpha = 0.0$ (no Perlin) with $\alpha = 0.4$. Even with 60% noise in the sample, the models outperform those trained with datasets without noise.

Note that the best setting, $\alpha = 0.4$, produces samples that look too "dirty", as can be seen in Fig. 11. Smaller settings will produce more realistic values, however, higher accuracy is obtained with this more saturated version.

Table 5. Results on GTSRB for SES datasets for Perlin noise blending.

$\alpha = 0.0$	$\alpha = 0.2$	$\alpha = 0.4$	$\alpha = 0.6$	$\alpha = 0.8$
98.60 ± 0.17	98.95 ± 0.12	**99.31** ± 0.12	99.06 ± 0.09	89.34 ± 0.23

Confetti noise was tested in combination with Perlin noise to evaluate its usefulness regarding accuracy, see Table 6.

Regarding datasets with only Confetti noise (no Perlin noise) vs. the base dataset (no Confetti nor Perlin Noise), the usage of Confetti noise brings consistent improvement.

Table 6. Combined Perlin and Confetti accuracy results for models trained with synthetic datasets for GTSRB. In the base datasets neither Perlin nor Confetti is applied.

datasets	C + P	P	C	base
SES	99.24 ± 0.12	$\mathbf{99.31 \pm 0.12}$	98.70 ± 0.17	98.60 ± 0.17
SER	$\mathbf{99.34 \pm 0.09}$	$\mathbf{99.34 \pm 0.08}$	99.24 ± 0.10	99.06 ± 0.10
SJS	$\mathbf{99.25 \pm 0.09}$	99.23 ± 0.08	98.70 ± 0.18	98.60 ± 0.17
SJR	$\mathbf{99.49 \pm 0.04}$	99.44 ± 0.07	99.09 ± 0.07	99.08 ± 0.17

Comparing the effect of Perlin noise vs. Confetti noise it is clear that the former has a significantly greater impact than the latter. This is to be expected as Confetti noise is only applied to smaller samples (20% of the dataset).

Combining both noise algorithms brings a somewhat unexpected result for the SES datasets. While all the other variations improve or at least maintain their accuracy when using both noise algorithms, there is a significant decrease in accuracy in the models trained with the SES datasets.

At the moment we don't have a justification for this behaviour, and further study is required to evaluate the Confetti noise impact when combined with Perlin noise. Nevertheless, as Confetti noise provides better results in the majority of cases, we will preserve it in all datasets for the sake of consistency with our previous work.

Results for all types of synthetic datasets in the three datasets used throughout this work can be found in Table 7. The results for all models trained with synthetic data are within less than 0.5% of the accuracy obtained with real data. This represents a clear step in bridging the gap between real and synthetic data.

Table 7. Test dataset accuracy for models trained with synthetic data. The accuracy obtained with the real dataset is presented inside parenthesis.

	SES	SER	SJS	SJR
GTSRB (99.64)	99.24 ± 0.12	99.34 ± 0.09	99.25 ± 0.09	$\mathbf{99.49 \pm 0.04}$
BTSC (99.30)	$\mathbf{99.12 \pm 0.04}$	98.86 ± 0.12	99.11 ± 0.09	98.92 ± 0.09
rMASTIF (99.72)	$\mathbf{99.47 \pm 0.09}$	99.27 ± 0.14	99.26 ± 0.17	99.37 ± 0.08

The best previously reported accuracy for a synthetic dataset was by Araar et al. in [1], presenting an accuracy for GTSRB of 97.83%. Our results, in all the variants clearly surpass previous results. Is it noteworthy to point out that our work is the only one which does not pursue photo-realism when generating samples, as can be seen in Fig. 13. This hints that pursuing photo-realism may not be a requirement, or even the best option.

Another interesting result is that, as opposed to the results reported by Araar et al. in [1], we managed to obtain very good results with solid colour backgrounds.

Considering our results, for both BTSC and rMASTIF the best result is obtained with the SES dataset. This is the most agnostic dataset, as it does not incorporate neither background or brightness information from real data. On the other hand, for GTSRB the best result was obtained almost in the opposite scenario: using both brightness information and real backgrounds.

When using brightness information for GTSRB the result is consistently better. This can be due to the fact that this dataset has the darkest samples on average and the narrowest brightness distribution curve, thereby benefiting from a more tailored brightness distribution. The fact that both BTSC and rMASTIF datasets offer better performance with solid backgrounds could be interpreted as a hint that the negative background dataset we are using is biased towards the German dataset.

5.2   Combining Real and Synthetic Data

Assuming real data is available, we can combine it with synthetic data in two ways: merging datasets and ensembling. This section describes both options.

Merging Real and Synthetic Data. To evaluate the benefits of merging both types of data, two synthetic versions were selected to be merged with real data: SES and SJS, i.e., solid background synthetic datasets with both brightness options. Based on the previously built datasets, we created 5 merged datasets for each brightness option.

Table 8. Average accuracy results for merged datasets.

	Real + SES	- Real + SJS
GTSRB	99.70 ± 0.04	**99.75 ± 0.02**
BTSC	99.36 ± 0.05	**99.40 ± 0.05**
rMASTIF	99.81 ± 0.04	**99.84 ± 0.07**

As expected, the results show a clear improvement over previous results, see Table 8. Results also show a slight advantage when using brightness information from the real dataset.

Ensembles. Ensembling is a known technique in deep learning when multiple models are available. To take advantage of the diversity of data available, our approach is to consider a synthetic dataset, a merged dataset, and the real dataset. This results in three trained models that we combine in an ensemble.

Considering that the trained merged models have solid colour backgrounds, the synthetic dataset will have real backgrounds. To increase diversity we will use the SER dataset, as it is more agnostic of the dataset than SJR since the latter includes brightness information from the training set.

Therefore, our ensemble has three models trained with the following datasets: SER, Real, and Merged (Real + SJS). This provides a diverse ensemble with both brightness and background options.

The ensemble was evaluated 5 times, selecting the ith model of each type to build the ith ensemble (Table 9).

Ensembling provides mixed results regarding accuracy. Both in BTSC and rMASTIF the results are worse compared to merging only. On a closer examination of the BTSC individual model results we can observe that there is a significant set of samples that are misclassified by most models. Hence, ensembling is unable to improve over the

Table 9. Average accuracy results for ensembles. Source [10].

GTSRB	BTSC	rMASTIF
99.82 ± 0.02	99.38 ± 0.02	99.79 ± 0.05

individual models accuracy. Note that the difference in percentage translates to a single sample as the test dataset is relatively small. In rMASTIF a similar situation was found.

On the other hand, for GTRSB the result is above the state of the art (99.81% from Haloi [4]), with the best ensemble for GTSRB achieved an accuracy of 99.85%. Although it may seem unfair to compare an ensemble to a single network, note that our input is only 32×32, compared to Haloi's 128×128. This implies that, although we are considering 3 models, it is likely that inference with our ensemble will be faster than with Haloi's model. Furthermore, our ensemble memory footprint is also smaller than Haloi's, see Table 4 for the number of parameters on both models.

5.3 Cross-Testing

In Sect. 2 we mentioned the existing intraclass variety that can be found when considering the same traffic sign for different countries. We also discussed how lighting and camera sensors can be a relevant issue when pursuing the best accuracy performance.

This test aims at exploring these issues. It is also an assessment on the generalization capabilities of the models trained with different datasets.

The test consists of evaluating a model trained with a dataset for a particular country in a dataset of a different country. For instance, models trained with GTSRB datasets will be evaluated on test datasets from Croatia and Belgium. Compared to [10], we have extended this test to include two more test countries: France and Sweden. Data for this datasets came from the ETSD, where we include all classes from France and Sweden that semantically overlapped classes in the three main datasets used previously in this work: GTSRB, BTSC, and rMASTIF. Furthermore, this test includes more classes than in [10]. Figure 14 show the classes that were considered and provides samples for the respective datasets. For this test we included all classes that have similar pictograms and fonts, although templates may vary slightly.

Note that for GTSRB, BTSC, and rMASTIF, we use only the test datasets for evaluation since this also provides for a direct comparison with the previous reported results. However, since the French and Swedish datasets are smaller and have not been used previously in this work, we opted to use the full dataset (training + test). Results are presented in Table 10.

Based on the results presented it is fair to say that using a dataset designed for a country in another country is not advisable, with the reported accuracies for models trained with real data falling as low as 60.66%. This low performance suggests that the test data does not belong to the distribution of the real data. Although, this test has been presented as a test relating to the usage of a dataset built for a country being used in another country, this test can also be seen as evaluating how different camera sensors, lighting, and national intraclass variation would impact accuracy. From this perspective, it confirms the difficulty in gathering a truly representative dataset, as discussed in Sect. 2.

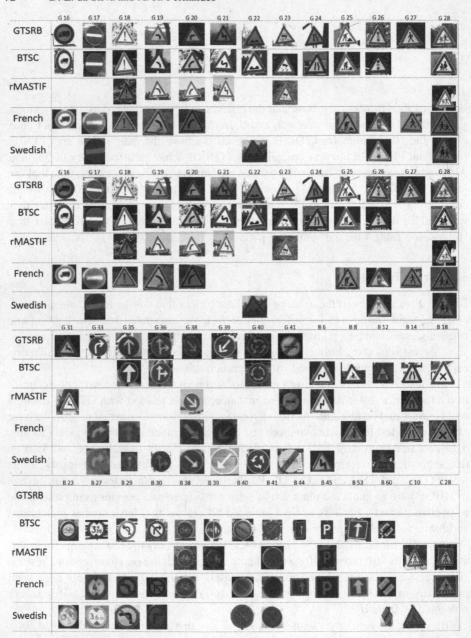

Fig. 14. Sample signs from classes with the same semantic meaning.

Yet, perhaps the most surprising result is the fact that synthetic datasets achieve higher accuracies in 9 out of 12 tests. In particular SES achieved the highest accuracy in 8 out of 12 scenarios. When considering the sum over each training dataset used, SES datasets provided the highest accuracies in all three cases, with real datasets offering

Table 10. Accuracy results for cross-testing. Each group describes the results obtained with models trained with the respective datasets; Rows indicate datasets used for evaluation purposes. Second column reports on the number of samples from the test datasets.

Test set	#	R	SER	SES
		Trained for GTSRB (Germany)		
Belgium	942	**97.98 ± 0.43**	97.47 ± 0.52	96.58 ± 1.06
Croatia	1067	95.99 ± 0.38	99.18 ± 0.35	**98.78 ± 0.32**
France	670	**81.94 ± 0.93**	80.24 ± 0.79	81.79 ± 1.33
Sweden	4258	60.66 ± 1.10	62.81 ± 1.64	**64.03 ± 2.14**
Total	6937	73.22 ± 0.76	74.64 ± 1.01	**75.51 ± 1.32**
		Trained for BTSC (Belgium)		
Croatia	1019	66.48 ± 0.62	84.26 ± 1.29	**84.91 ± 1.60**
German	5699	83.98 ± 0.97	**95.16 ± 0.39**	93.69 ± 0.89
France	1100	77.95 ± 0.44	79.97 ± 0.90	**79.69 ± 0.88**
Sweden	2395	**73.53 ± 0.81**	65.57 ± 1.89	70.81 ± 1.22
Total	10213	79.13 ± 0.61	85.47 ± 0.62	**85.94 ± 0.74**
		Trained for rMASTIF (Croatia)		
Belgium	1169	76.92 ± 1.71	84.36 ± 1.01	**85.37 ± 0.09**
German	7109	91.65 ± 0.79	96.32 ± 0.72	**97.21 ± 0.59**
France	685	80.93 ± 0.77	77.69 ± 1.21	**81.02 ± 1.27**
Sweden	3601	70.83 ± 0.97	75.11 ± 1.15	**77.97 ± 1.26**
Total	12564	83.73 ± 0.42	88.11 ± 0.54	**89.71 ± 0.26**

the lowest accuracies. The predominance of SES over SER, the only difference being in the backgrounds, again hints that the usage of real backgrounds has the potential to introduce an undesirable bias. SES is the most agnostic dataset, not including lighting information from the respective real counterpart dataset. Together with the poor results obtained with the real datasets, this suggests that these real datasets are in fact not representative of the global population of traffic sign imagery.

Not only does SES provide the best results, but the differences towards the results obtained with real data can be very significant. The worst case is when comparing models trained for BTSC and tested on the Croatia test set, where the difference between accuracies reaches 18.43%. On the other hand, in the 3 scenarios where R beats SES the differences are significantly smaller: 1.4%, 0.15%, and 2.72%.

This difference between results obtained synthetic and real datasets indicates that the former may provide better generalization than the latter. This is particularly significant as SES is the most agnostic dataset.

5.4 Unleashing Synthetic Datasets

Up until now we gathered templates for each dataset observing only the training dataset. The set of templates for each class may therefore not be fully representative. This section unleashes the synthetic datasets in the sense that we are now free to examine the test datasets to seek for new templates.

Upon this exploration of the test sets it became clear that both in BTSC and GTSRB there are multiple signs whose templates are not found in the previous datasets. This is to be expected in real scenarios as discussed in Sect. 2, as gathering real samples for all pictogram or font variations is an immensely time and resource consuming task.

This section reports on this exploration for two datasets: BTSC and GTSRB.

BTSC. In the Belgium dataset we found a set of six images from the test set that are misclassified by the majority of models, both trained with synthetic and real data. This problem was identified when we analysed the ensemble results. Figure 15 shows these samples. These samples belong to the same class, class 45, however, in the training set there are no samples with such templates. Figure 16 presents the initial templates considered, and the new templates added for this test.

Fig. 15. BTSC - Set of images misclassified by the majority of the models, both trained on real and synthetic data. Source [10].

Repeating the previous test with the SES dataset we found that not only we got 100% accuracy for class 45 but also that there were no adversarial effects on the other classes. Adding this templates we got an average accuracy of 99.31%, surpassing the results we got with the real dataset.

Although the difference to the accuracy obtained with the real dataset is relatively small (0.01%) the synthetic dataset behaves considerably better than its real counterpart when considering the average accuracy obtained during training, as seen in Fig. 17.

Fig. 16. BTSC - left: templates from the training set; right: new templates from the test set (note: the text was added just for the template and does not correspond to any real sign). Source [10].

As a final test we merged one of these datasets with the real BTSC dataset, achieving 99.76% accuracy, with only 6 misclassified images out of 2520. This result surpasses the current state of the art result of 99.72% reported in [13], with inputs that have a quarter of the pixels and a model with a lighter architecture. The best epoch achieved an accuracy of 99.84%.

GTSRB. In this dataset misclassifications are spread over several classes. Examining the templates from the test set we ended adding the templates in Fig. 18.

These templates represent variations that are missing from the training set. Technically, the third sign in Fig. 18 is not a traffic sign, but it is a common combination present in both the training and test sets.

Repeating the previous tests with the unleashed dataset we noticed a decrease in the number of misclassified samples in classes where templates were added without

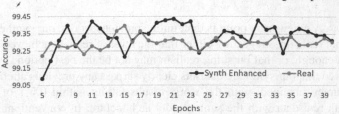

Fig. 17. BTSC - accuracy for the test set. Comparison between models trained with the enhanced synthetic training sets vs models trained with real data. Source [10].

Fig. 18. Added templates for GTSRB.

relevant adversarial side effects. Table 11 provides a comparison between the previously obtained accuracies and the results for the unleashed datasets, showing a significant improvement in all synthetic models.

Table 11. Results for unleashed GTSRB datasets. Source [10].

	Prev Results	Unleashed
SES	99.24 ± 0.12	**99.41 ± 0.08**
SER	99.34 ± 0.09	**99.40 ± 0.10**
SJS	99.25 ± 0.09	**99.50 ± 0.09**
SJR	99.49 ± 0.04	**99.57 ± 0.04**

As previously done for BTSC, we also tested merging real data with one of the SES and SJR models. The accuracy obtained was 99.80% for Real + SES, and 99.79% for Real + SJR. These are marginally below the state of the art results reported in [4] (99.81%). As noted previously in Sect. 5.1, our models are much lighter than Haloi's due to the difference in image input size (32×32 vs. 128×128).

6 Conclusion

A new proposal for synthetic data generation for traffic sign repositories was presented. Two operators were introduced: Perlin and Confetti noise. While Perlin noise provides a clear advantage Confetti noise still requires some further study to evaluate its usefulness. We also explored taking advantage of the brightness distribution of real data when available. While using real data distributions for some variable can provide an increase in accuracy in some scenarios it also may introduce undesirable bias to the dataset. Finally, we also explored solid colour vs. real imagery backgrounds. Again, while real background imagery may provide some benefits in some scenarios, it is also a potential source of bias. A noteworthy point is that, unlike previous works, our synthetic samples are clearly not realistic, suggesting either that the level of realism achieved in previous works is not enough, or that pursuing realism may not be the best option.

We showed that our synthetic datasets clearly surpass any previous attempts regarding the accuracy obtained, with the worst results falling within 0.5% of the results obtained with real data with the same model architecture. In conventional tests such as merging and ensembling we surpassed state of the art results in three public traffic sign datasets, clearly showing the potential of synthetic data.

While surpassing state of the art results is always a rewarding, we strongly believe that the most relevant results are those when synthetic data was unleashed, and the cross-testing experiments.

Unleashing the synthetic dataset provided excellent results at a minimal cost. All that is required is to retrain the model. The set of templates for unleashed datasets is the natural set of templates when no real data is available. In this scenario all templates for older versions should be collected to achieve the full potential of a trained model.

Cross-testing was presented as in inter-country test, but, as discussed, it can also be seen as a broader national test with a very diverse test set. We obtained an interesting result with the most agnostic dataset, without real backgrounds and without using the brightness distribution of real data, clearly surpassing the results obtained with real data. This is a clear indication of the generalization potential when using synthetic data.

We strongly believe that our work shows the potential of synthetic data in the domain of traffic sign repositories. Nevertheless, there is still work to be done. The pipeline can be fine tuned, and new operators can be explored to deal with shadows, rain, fog, and night time, amongst other common occurrences in traffic sign imagery.

Acknowledgements. This work has been supported by FCT - Fundação para a Ciência e Tecnologia within the RD Units Project Scope: UIDB/00319/2020.

References

1. Araar, O., Amamra, A., Abdeldaim, A., Vitanov, I.: Traffic sign recognition using a synthetic data training approach. Int. J. Artif. Intell. Tools **29**, 2050013 (2020). https://doi.org/10.1142/S021821302050013X
2. Grigorescu, C., Petkov, N.: Distance sets for shape filters and shape recognition. IEEE Trans. Image Process. **12**(10), 1274–1286 (2003). https://doi.org/10.1109/TIP.2003.816010
3. Gámez Serna, C., Ruichek, Y.: Classification of traffic signs: the European dataset. IEEE Access **6**, 78136–78148 (2018). https://doi.org/10.1109/ACCESS.2018.2884826
4. Haloi, M.: Traffic sign classification using deep inception based convolutional networks. arXiv abs/1511.02992 (2015)
5. Horn, D., Houben, S.: Fully automated traffic sign substitution in real-world images for large-scale data augmentation. In: 2020 IEEE Intelligent Vehicles Symposium (IV), pp. 465–471 (2020). https://doi.org/10.1109/IV47402.2020.9304547
6. Horn, D., Janssen, L., Houben, S.: Automated selection of high-quality synthetic images for data-driven machine learning: a study on traffic signs. In: 2021 IEEE Intelligent Vehicles Symposium (IV), pp. 832–837 (2021). https://doi.org/10.1109/IV48863.2021.9575337
7. Jurišić, F., Filković, I., Kalafatić, Z.: Multiple-dataset traffic sign classification with onecnn. In: 2015 3rd IAPR Asian Conference on Pattern Recognition (ACPR), pp. 614–618 (2015). https://doi.org/10.1109/ACPR.2015.7486576

8. Larsson, F., Felsberg, M.: Using Fourier descriptors and spatial models for traffic sign recognition. In: Heyden, A., Kahl, F. (eds.) SCIA 2011. LNCS, vol. 6688, pp. 238–249. Springer, Heidelberg (2011). https://doi.org/10.1007/978-3-642-21227-7_23

9. Liu, Y.T., Chen, R.C., Dewi, C.: Generate realistic traffic sign image using deep convolutional generative adversarial networks. In: 2021 IEEE Conference on Dependable and Secure Computing (DSC), pp. 1–6 (2021). https://doi.org/10.1109/DSC49826.2021.9346266

10. Lopes da Silva, D., Ramires Fernandes, A.: Bridging the gap between real and synthetic traffic sign repositories. In: Proceedings of the 3rd International Conference on Deep Learning Theory and Applications - DeLTA, pp. 44–54. INSTICC, SciTePress (2022). https://doi.org/10.5220/0011301100003277

11. Luo, H., Kong, Q., Wu, F.: Traffic sign image synthesis with generative adversarial networks. In: 2018 24th International Conference on Pattern Recognition (ICPR), pp. 2540–2545 (2018)

12. Luo, J., Wang, Z.: A low latency traffic sign detection model with an automatic data labeling pipeline. Neural Comput. Appl. 1–14 (2022)

13. Mahmoud, M.A.B., Guo, P.: A novel method for traffic sign recognition based on DCGAN and MLP with PILAE algorithm. IEEE Access 7, 74602–74611 (2019). https://doi.org/10.1109/ACCESS.2019.2919125

14. Paparoditis, N., et al.: Stereopolis II: a multi-purpose and multi-sensor 3D mobile mapping system for street visualisation and 3D metrology. Revue Française de Photogrammétrie et de Télédétection (200), 69–79 (2014). https://doi.org/10.52638/rfpt.2012.63

15. Perlin, K.: An image synthesizer. SIGGRAPH Comput. Graph. 19(3), 287–296 (1985). https://doi.org/10.1145/325165.325247

16. Saha, S., Kamran, S.A., Sabbir, A.S.: Total recall: understanding traffic signs using deep hierarchical convolutional neural networks. CoRR abs/1808.10524 (2018). https://arxiv.org/abs/1808.10524

17. Spata, D., Horn, D., Houben, S.: Generation of natural traffic sign images using domain translation with cycle-consistent generative adversarial networks. In: 2019 IEEE Intelligent Vehicles Symposium (IV), pp. 702–708 (2019)

18. Stallkamp, J., Schlipsing, M., Salmen, J., Igel, C.: Man vs. computer: benchmarking machine learning algorithms for traffic sign recognition. Neural Netw. (2012). https://doi.org/10.1016/j.neunet.2012.02.016. https://www.sciencedirect.com/science/article/pii/S0893608012000457

19. Stergiou, A., Kalliatakis, G., Chrysoulas, C.: Traffic sign recognition based on synthesised training data. Big Data Cogn. Comput. 2(3) (2018). https://doi.org/10.3390/bdcc2030019. https://www.mdpi.com/2504-2289/2/3/19

20. Timofte, R., Zimmermann, K., Gool, L.V.: Multi-view traffic sign detection, recognition, and 3D localisation. In: 2009 Workshop on Applications of Computer Vision (WACV), pp. 1–8 (2009). https://doi.org/10.1109/WACV.2009.5403121

21. Youssef, A., Albani, D., Nardi, D., Bloisi, D.D.: Fast traffic sign recognition using color segmentation and deep convolutional networks. In: Blanc-Talon, J., Distante, C., Philips, W., Popescu, D., Scheunders, P. (eds.) ACIVS 2016. LNCS, vol. 10016, pp. 205–216. Springer, Cham (2016). https://doi.org/10.1007/978-3-319-48680-2_19

22. Zhu, Z., Liang, D., Zhang, S., Huang, X., Li, B., Hu, S.: Traffic-sign detection and classification in the wild. In: 2016 IEEE Conference on Computer Vision and Pattern Recognition (CVPR), pp. 2110–2118 (2016)

23. Šegvic, S., et al.: A computer vision assisted geoinformation inventory for traffic infrastructure. In: 13th International IEEE Conference on Intelligent Transportation Systems, pp. 66–73 (2010). https://doi.org/10.1109/ITSC.2010.5624979

Convolutional Neural Networks for Structural Damage Localization on Digital Twins

Marco Parola[1]([✉]) [iD], Federico A. Galatolo[1] [iD], Matteo Torzoni[2] [iD],
and Mario G. C. A. Cimino[1] [iD]

[1] Department of Information Engineering, University of Pisa, Largo L. Lazzarino 1,
Pisa, Italy
{marco.parola,federico.galatolo,mario.cimino}@ing.unipi.it
[2] Department of Civil and Environmental Engineering, Politecnico di Milano,
Piazza L. da Vinci 32, Milano, Italy
matteo.torzoni@polimi.it

Abstract. Structural health monitoring (SHM) using IoT sensor devices plays a crucial role in the preservation of civil structures. SHM aims at performing an accurate damage diagnosis of a structure, that consists of identifying, localizing, and quantify the condition of any significant damage, to keep track of the relevant structural integrity. Deep learning (DL) architectures have been progressively introduced to enhance vibration-based SHM analyses: supervised DL approaches are integrated into SHM systems because they can provide very detailed information about the nature of damage compared to unsupervised DL approaches. The main drawback of supervised approach is the need for human intervention to appropriately label data describing the nature of damage, considering that in the SHM context, providing labeled data requires advanced expertise and a lot of time. To overcome this limitation, a key solution is a digital twin relying on physics-based numerical models to reproduce the structural response in terms of the vibration recordings provided by the sensor devices during a specific events to be monitored. This work presents a comprehensive methodology to carry out the damage localization task by exploiting a convolutional neural network (CNN) and parametric model order reduction (MOR) techniques to reduce the computational burden associated with the construction of the dataset on which the CNN is trained. Experimental results related to a pilot application involving a sample structure, show the potential of the proposed solution and the reusability of the trained system in presence of different loading scenarios.

Keywords: Convolutional neural network · IoT · Digital twin · Structural health monitoring · Damage localization

1 Introduction and Background

Civil structures, whether buildings, bridges, oil and gas pipelines, are subject to several external actions and sources of degradation that might compromise their structural performance. This can happen due to a faulty construction process, lack of quality control, or unexpected loadings, environmental actions and natural hazards such as earthquakes. In order to keep track of the structural health, and to quickly react before a major damage occurs, autonomus damage identification systems provide a suitable framework to perform systematic diagnostic and prognostic activities, ultimately allowing for timely maintenance actions with a direct impact on reducing the operating costs.

In the last years, increasingly sophisticated structural health monitoring (SHM) systems have been developed to keep under control the structural health state, through the implementation of different levels of damage identification, such as detection, localization and quantification, possibly with a quantification of the impact of relevant environmental effects, see e.g. [19,21]. A SHM system is usually comprised by different components: a sensor network deployed to collect vibration and environmental data; a data transmission unit; a storage unit; a SHM data analysis software. The outcome of the monitoring procedure is then displayed to the user via reports or web platforms.

Recently, many vibration-based SHM strategies relying upon deep learning (DL) architectures have been proposed to address different SHM tasks, see e.g. [15–17], by exploiting their capability to automatize the selection and extraction of damage-sensitive features, that traditional algorithms would fail to detect [20]. Focusing on these data-driven strategies, the SHM problem can be addressed either in a supervised or an unsupervised way. The former exploits labeled input-output pairs, with vibration structural response data being the inputs and the values of the sought damage parameters being the corresponding outputs. On the other hand, unsupervised algorithms are often adopted to discover damage-sensitive patterns in the input data, without the need of providing a corresponding output label. The implementation of an unsupervised damage detection strategy usually involves two phases: (i) the identification of stable key parameters reflecting the undamaged structural health state by means of appropriate behavioral model of the structure; (ii) the detection of a persistent variation in such parameters over time, by relying upon the underlying idea that when the structure suffers damage, a deviation from the reference condition can be observed in terms of vibration response [5].

Since unsupervised learning methods can only be effective in detecting the presence of a structural damage, without allowing to obtain clear and explicit information about location, severity and type of damage, a supervised learning strategy is adopted in this paper. However, dealing with civil structures, experimental labeled data referred to the possible damage states can not be obtained in practice. To overcome this limitation, a key solution is provided by the digital twin (DT) paradigm, and in particular by the physics-based numerical models comprising the DT of the structure to be monitored. Indeed, the latter enable to systematically simulate the vibration recordings provided by IoT devices for specific damage and operational conditions [2].

A DT is built upon three components: a physical asset in the real-world (physical twin), a digital model of the structure in a computerized environment, and the integration of data and information that tie the physical entities and their virtual representation together [7]. For a successful DT implementation, it is crucial to properly identify all the involved physical entities and processes in the real-world and their digital counterparts, as well as the interconnection between them, in terms of the exchanged data. The process of implementing a DT is called digital transformation, and in this work it is (partially) achieved by means of a physics-based numerical model of the monitored structure relying on the finite element (FE) method and of a DL-based model for damage identification.

As the number of involved degrees of freedom increases, the computational cost associated to the solution of a FE model grows, and the assembly of synthetic datasets accounting for different input parameters easily becomes prohibitive [13]. To this aim, a reduced-order modeling strategy for parametrized systems is adopted by relying on the reduced basis method [12] in order to set a cheaper, yet accurate, reduced-order model (ROM), ultimately allowing to speed up data generation phase required to train the DL model in a supervised fashion.

This work deepens and extends a former research activity presented in [11], by proposing a comprehensive approach to solve the damage localization task. The obtained results testify the capabilities of the proposed approach to perform real-time damage localization, as well as the effectiveness of such diagnostic framework against operational variability and measurement noise. In addition, the beneficial effect of implementing a hyperparameter optimization strategy is also considered, as reported to yield a fair improvement in the damage localization performance.

The reminder of the paper organized as follows: the methods for developing the digital twin, and involving the generation of synthetic datasets and the use of DL-based architectures for SHM purposes are described in Sect. 2; the application of the methodology to the case study of a two-story shear building is discussed in Sect. 4; conclusions and future developments are finally reported in Sect. 5.

2   Methodology and Methods

The proposed methodology is described in the following. Specifically: in Sect. 2.1, we detail the numerical models comprising the DT, adopted to populate the synthetic dataset; in Sect. 2.2, we frame the damage localization task as a classification problem handling a set of predefined damage scenarios by means of a convolutional neural network (CNN).

2.1   Digital Twin Design

Physics-Based Numerical Model. The virtual representation of the structure to be monitored is obtained by relying upon a high-fidelity full-order model (FOM), describing its dynamic response under the applied loadings, according to the Newton's second law of motion and under the assumption of a linearized kinematic. By modeling the structure as a linear-elastic continuum and by space-discretizing the governing equation by means of a FE mesh, its dynamic response is described by the following semi-discretized form of the elasto-dynamic problem:

$$\begin{cases} \mathbf{M}\ddot{\mathbf{d}}(t) + \mathbf{C}\dot{\mathbf{d}}(t) + \mathbf{K}(\varDelta, l)\mathbf{d}(t) = \mathbf{f}(t, \boldsymbol{\eta}) \ , \ t \in (0, T) \\ \mathbf{d}(0) = \mathbf{d}_0 \\ \dot{\mathbf{d}}(0) = \dot{\mathbf{d}}_0 \ , \end{cases} \tag{1}$$

where: $t \in (0, T)$ denotes time; $\mathbf{d}(t), \dot{\mathbf{d}}(t), \ddot{\mathbf{d}} \in \mathbb{R}^{\mathscr{M}}$ are the vectors of nodal displacements, velocities and accelerations, respectively; $\mathscr{M}$ is the number of degrees of freedom (dofs); $\mathbf{M} \in \mathbb{R}^{\mathscr{M} \times \mathscr{M}}$ is the mass matrix; $\mathbf{C} \in \mathbb{R}^{\mathscr{M} \times \mathscr{M}}$ is the damping matrix, assembled according to the Rayleigh's model; $\mathbf{K}(\varDelta, l) \in^{\mathscr{M} \times \mathscr{M}}$ is the stiffness matrix, with $\varDelta$ and l the parameters providing its dependence on damage as specified below; $f(t, \boldsymbol{\eta}) \in \mathbb{R}^{\mathscr{M}}$ is the vector of nodal forces associated to the operational conditions ruled by means of N_η parameters through the vector $\boldsymbol{\eta} \in \mathbb{R}^{N_\eta}$; $\mathbf{d}_0 \in \mathbb{R}^{\mathscr{M}}$ and $\dot{\mathbf{d}}_0 \in \mathbb{R}^{\mathscr{M}}$ are the initial conditions at $t = 0$, in terms of nodal displacements and velocities, respectively. The relevant parametric input space is assumed to display a uniform probability distribution for each parameter.

As typically done in simulation-based SHM, damage is modeled as a selective degradation of the material stiffness of amplitude $l \in \mathbb{R}$, taking place within the pre-designated region labeled by $\varDelta \in \{\varDelta_0, \ldots, \varDelta_{N_d}\}$, with $\varDelta_0$ identifying the damage-free baseline and all the others being referred to specific damage scenarios undergone by the structure among a set of predefined N_d damage states. These latter are defined on the basis of structural response, loading conditions, and aging processes of materials. In this work, l is not considered part of the label, as only the localization of damage is addressed.

Dataset Generation. The generation of the training instances is carried out by advancing in time the solution of the physics-based model of the structure using the Newmark time integration scheme. Either nodal displacements or accelerations recordings $\boldsymbol{\delta}_i = \boldsymbol{\delta}_i(\varDelta, l, \boldsymbol{\eta}) \in \mathbb{R}^L$ in $(0, T)$, each including L measurements, are collected at N_s predefined locations where sensing devices are supposed to be installed, with $i = 1, \ldots, N_s$. The measurements are acquired with a sampling frequency f and for an observation time window $(0, T)$, short enough to assume constant operational and damage conditions, such that $T = (L - 1)/f$. The training set $\mathbf{D} \in \mathbb{R}^{L \times N_s \times N_o}$ is then built from the assembly of N_o instances, each one shaped as a multivariate time series comprised by N_s arrays of L measurements and obtained by sampling the parametric input space of the numerical model via latin hypercube rule.

In order to obtain a high quality dataset $\mathbf{D}$ to train the DL model, the number of required instances may be extremely high, thus making the computational cost associated to the data generation process potentially very high. To this aim, the FOM is replaced by a cheaper, yet accurate, projection-based ROM by relying on the reduced basis method [12], following the same strategy adopted in [13,16,18].

By relying upon the proper orthogonal decomposition (POD)-Galerkin approach, the solution to Problem (1) is approximated, in terms of displacements, as $\mathbf{d}(t, \varDelta, l, \boldsymbol{\eta}) \approx \mathbf{W}\mathbf{d}_R(t, \varDelta, l, \boldsymbol{\eta})$, which is a linear combination of $\mathscr{M}_R \ll \mathscr{M}$ basis functions $\mathbf{w}_r \in \mathbb{R}^{\mathscr{M}}$, $r = 1, \ldots, \mathscr{M}_R$, gathered in the projection matrix

$\mathbf{W} = [\mathbf{w}_1, \dots, \mathbf{w}_R] \in \mathbb{R}^{\mathcal{M} \times \mathcal{M}_R}$, with $\mathbf{d}_R(t, \Delta, l, \boldsymbol{\eta}) \in \mathbb{R}^{\mathcal{M}_R}$ being the vector of unknown POD-coefficients

By enforcing the orthogonality between the residual and the subspace spanned by the first $\mathcal{M}_R$ POD-modes through a Galerkin projection, the following $\mathcal{M}_R$-dimensional dynamical system is obtained:

$$\begin{cases} \mathbf{M}_R \ddot{\mathbf{d}}(t) + \mathbf{C}_R \dot{\mathbf{d}}(t) + \mathbf{K}_R(\Delta, l) \mathbf{d}_R(t) = \mathbf{f}_R(t, \boldsymbol{\eta}) \, , \, t \in (0, T) \\ \mathbf{d}_R(0) = \mathbf{W}^\top \mathbf{d}_0 \\ \dot{\mathbf{d}}_R(0) = \mathbf{W}^\top \dot{\mathbf{d}}_0 \, , \end{cases} \tag{2}$$

Here, the reduced arrays play the same role of their HF counterparts, yet with dimension ruled by $\mathcal{M}_R$ instead of $\mathcal{M}$, according to:

$$\mathbf{M}_R \equiv \mathbf{W}^\top \mathbf{M} \mathbf{W} \, , \ \mathbf{C}_R \equiv \mathbf{W}^\top \mathbf{C} \mathbf{W} \, , \ \mathbf{K}_R \equiv \mathbf{W}^\top \mathbf{K} \mathbf{W} \, , \ \mathbf{f}_R \equiv \mathbf{W}^\top \mathbf{f}. \tag{3}$$

The approximated solution is then recovered by back-projecting the ROM solution, via $\mathbf{d}(t) \approx \mathbf{W} \mathbf{d}_R(t)$, or $\ddot{\mathbf{d}}(t) \approx \mathbf{W} \ddot{\mathbf{d}}_R(t)$ depending on the handled measurements.

The projection matrix $\mathbf{W}$ is obtained by performing a singular value decomposition of a snapshot matrix $\mathbf{S} = [\mathbf{d}_1, \dots, \mathbf{d}_{\mathcal{S}}] \in \mathbb{R}^{\mathcal{M} \times \mathcal{S}}$, assembled from $\mathcal{S}$ snapshots of the FOM, namely solutions in terms of time histories of nodal displacements, obtained for different values of the parameters, as

$$\mathbf{S} = \mathbf{P} \boldsymbol{\Sigma} \mathbf{Z}^\top \, , \tag{4}$$

where: $\mathbf{P} = [\mathbf{p}_1, \dots, \mathbf{p}_{\mathcal{M}}] \in \mathbb{R}^{\mathcal{M} \times \mathcal{M}}$ is an orthogonal matrix, whose columns are the left singular vectors of $\mathbf{S}$; $\boldsymbol{\Sigma} \in \mathbb{R}^{\mathcal{M} \times \mathcal{S}}$ is a pseudo-diagonal matrix collecting the singular values of $\mathbf{S}$, arranged so that $\sigma_1 \geq \sigma_2 \geq \cdots \geq \sigma_{\mathcal{R}} \geq 0$, $\mathcal{R} = min(\mathcal{S}, \mathcal{M})$ being the rank of $\mathbf{S}$; $\mathbf{Z} = [\mathbf{z}_1, \dots, \mathbf{z}_{\mathcal{S}}] \in \mathbb{R}^{\mathcal{S} \times \mathcal{S}}$ is an orthogonal matrix, whose columns are the right singular vectors of $\mathbf{S}$.

The ROM order $\mathcal{M}_R$ is set by adopting a standard energy-content criterion by prescribing a tolerance ϵ on the fraction of energy content to be disregarded in the approximation, according to:

$$\frac{\sum_{m=1}^{\mathcal{M}_R} (\sigma_m)^2}{\sum_{m=1}^{\mathcal{R}} (\sigma_m)^2} \geq 1 - \epsilon^2 \, , \tag{5}$$

that is the energy retained by the last $\mathcal{R} - \mathcal{M}_R$ POD-modes is equal or smaller than ϵ^2.

2.2　Deep Learning for the Damage Localization

In this paper, we propose the use of one-dimensional (1D) CNNs to solve the damage localization task, framed as a multiclass classification problem. A classification task involves the prediction of an output class label based on a given input. In this case, the output labels to be predicted identify a set of predefined damage scenarios, each referring to a different damage location.

Originally developed within the computer vision community, convolutional layers have quickly become a first choice to solve several problems, outperforming alternative methods [6,8]. They feature good relational inductive biases such as the locality and translational equivariance (parameter sharing) of convolutional kernels, which prove highly effective to analyze multivariate time series while improving the relevant computational efficiency. In a convolutional layer, the kernel filters feature a charcteristic size controlling the width of the local receptive field on its input. Each convolutional layer simultaneously applies multiple kernel filters throughout its input, resulting in multiple activation maps, called feature maps, each one providing the location and strength of the relevant convolutional kernel in the input.

In a deep learning framework, a multi-label classification task can be addressed by prescribing number of computational neurons in the last layer of the adopted neural network equal to the number of possible target labels. The resulting architecture is typically trained to solve the underlying classification task by minimizing the categorical cross-entropy between the predicted and target label classes:

$$H(\boldsymbol{b}, \widehat{\boldsymbol{b}}) = - \sum_{j=0}^{N_d} b_j \, log\widehat{b}_j \, , \qquad (6)$$

where $\boldsymbol{b} = \{b_0, \ldots, b_{N_d}\}^\top \in \mathbb{B}^{N_d}$ and $\widehat{\boldsymbol{b}} = \{\widehat{b}_0, \ldots, \widehat{b}_{N_d}\}^\top \in \mathbb{R}^{N_d}$ are two vectors gathering the Boolean indexes b_j, whose value is 1 if the target class for the current instance is j and 0 otherwise, and the confidence levels $\widehat{b}_j$ by which the current instance is assigned to the j-th damage class, respectively.

To evaluate the performance of the adopted DL model against the considered multi-class classification problem, the accuracy, precision, and recall indicators are used, as follows:

$$\text{Accuracy} = \frac{\text{TP} + \text{TN}}{\text{TP} + \text{TN} + \text{FP} + \text{FN}} \, ; \qquad (7)$$

$$\text{Precision} = \frac{\text{TP}}{\text{TP} + \text{FP}} \, ; \qquad (8)$$

$$\text{Recall} = \frac{\text{TP}}{\text{TP} + \text{FN}} \, . \qquad (9)$$

Herein, TP denotes the amount of true positives, FP denotes the amount of false positives, TN denotes the amount of true negatives, FN the amount of false negatives.

3   Case Study

The damage location capabilities of the proposed methodology are assessed with reference to the virtual health monitoring of the building reported in Fig. 1a.

Within a digital twin perspective, this physical asset is equipped with a network of sensors, for instance using commercial IoT devices as the one reported in Fig. 1b. This latter is a mono-axial wireless device, useful to acquire displacement measurements with an accuracy of 0.01 mm[1] A crucial aspect to take into account in the practice is the synchronization between IoT devices, which is a critical requirement for system operation. This latter is not a zero-cost process and different protocols can be adopted to meet the prescribed requirements, depending on the system type [22].

In the following, the corresponding digital twin is developed in order to suitably perform real-time damage detection and localization under the action of seismic loads. To provide a faithful virtual description of the considered framework, the digital transformation process is carried out by means of a DT comprised by three components: (i) a physics-based model of the structure to be monitored, either the FOM governed by Problem (1) or the corresponding reduced-order representation described by Problem (2); (ii) the generation of site-specific accelerograms, compatible with pseudo-real seismic loads and exploited to force the structural model; (iii) the extraction of relevant dofs recordings, in terms of multi-variate time series to mimic the deployed monitoring system, and the contamination of these signals by adding an independent, identically distributed Gaussian noise to allow for measurement noise.

(a)   (b)

Fig. 1. Pilot example: (a) physical twin to be monitored; (b) exemplary IoT device suitable for dynamic monitoring.

The building to be monitored is modeled as a two-dimensional frame, see Fig. 2, adopting a plane stress formulation with an out of plane thickness of 0.1 m. The structure is assumed to be made of concrete, with mechanical properties: Young's modulus $E = 30$ GPa, Poisson's ratio $\nu = 0.2$, density $\rho = 2500$ kg/m^3.

The structure is excited by seismic loads simulated by means of ground motion prediction equations adapted from [9,14] and allowing to generate spectrum-compatible accelerograms as a function of: local magnitude $Q \in$

---

[1] Deck, Dynamic Displacement Sensor. Move Srl, Italy. https://www.movesolutions. it/deck/.

[4.6, 5.3], epicentral distance $R \in [80, 100]$ km, and site geology; with the previous notation used to specify the ranges in which Q and R can take value, implicitly denoting that a uniform probability distribution is adopted to describe them, while having considered a rocky condition as site geology.

Structural displacement time histories $\boldsymbol{\delta}_i = \boldsymbol{\delta}_i(\Delta, l, \boldsymbol{\eta}) \in \mathbb{R}^L$, with $i = 1, \ldots, N_s$, are reocorded in $(0, T)$ from $N_s = 6$ dofs arranged as depicted in Fig. 2. Recordings are provided for a duration ($T = 70$ s) with an acquisition frequency of $f = 25$ Hz, thus consisting of $L = 1751$ measurements each. The FOM in Eq. (1) is obtained from a FE discretization using linear tetrahedral elements and resulting in $\mathscr{M} = 4326$ dofs. The damping matrix is assembled according to the Rayleigh's model to account for a 5% damping ratio on the first four structural modes. The stiffness matrix $\mathbf{K}(\Delta, l)$ is parameterized to account for $N_d = 9$ damage scenarios, simulated by reducing the material stiffness within the corresponding subdomain labeled by $\Delta \in \{\Delta_0, \ldots, \Delta_{N_d}\}$ and highlighted in dark grey in Fig. 2, with Δ_0 identifying the undamaged case and all the others being referred to specific damage state undergone by the structure, as described in Table 1. The damage level $l \in [5\%, 25\%]$, representing the amplitude of the stiffness degradation is held constant within the time instance $(0, T)$.

The projection basis $\mathbf{W}$ ruling the ROM in Eq. (2) is instead computed from a snapshot matrix $\mathbf{S}$ comprising $\mathscr{S} = 630,360$ snapshots, obtained through 360 evaluations of the FOM for different values of the input parameters sampled via

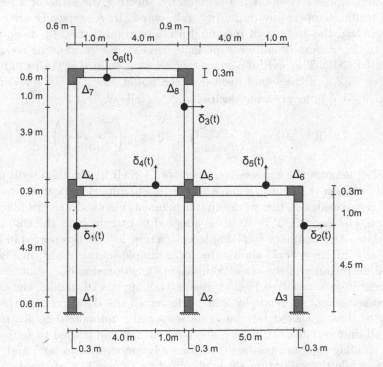

Fig. 2. Physics-based digital twin of the monitored structure, with details of the loading conditions and synthetic recordings related to displacements $\delta_1(t), \ldots, \delta_6(t)$ and target damage locations associated to $\Delta_1, \ldots, \Delta_8$.

Table 1. Considered damage scenarios.

Class label	Damage Location
Δ_0	Undamaged condition
Δ_1	Ground floor - left
Δ_2	Ground floor - mid
Δ_3	Ground floor - right
Δ_4	First floor - left
Δ_5	First floor - mid
Δ_6	First floor - right
Δ_7	Roof - left
Δ_8	Roof - mid

latin hypercube rule. By prescribing a tolerance $\epsilon = 10^{-4}$, the order of the ROM is set to $\mathcal{M}_R = 31$, in place of the original $\mathcal{M} = 4326$ dofs. The population of the dataset $\mathbf{D} \in \mathbb{R}^{L \times N_s \times N_o}$ is carried out by systematically evaluating the ROM for $N_o = 9999$ instances at varying input parameters values.

In this work, the signals are corrupted by assuming an additive Gaussian noise uncorrelated in time, to represent measurement noise and those environmental and ambient components potentially affecting the structural response, such as traffic, temperature, humidity, rain, wind [1]. As typically done in signal processing, the amount of meaningful information carried by a signal with respect to the amount of noise components is measured by adopting a signal-to-noise ratio (SNR). The SNR of a generic signal $\boldsymbol{\delta}^S$ is defined as the ratio between the power $\mathcal{P}_{\text{signal}}$ of the signal itself over the power $\mathcal{P}_{\text{noise}}$ of the relevant noise components $\boldsymbol{\delta}^N$, in logarithmic decibel scale as follows:

$$\text{SNR} = 10 \log_{10} \left(\frac{\mathcal{P}_{\text{signal}}}{\mathcal{P}_{\text{noise}}} \right) = 10 \log_{10} \left(\frac{\mathbb{E}[(\boldsymbol{\delta}^S)^2]}{\mathbb{E}[(\boldsymbol{\delta}^N)^2]} \right), \quad (10)$$

where $\mathbb{E}[\cdot]$ denotes the expectation operator. A SNR higher than 0 dB denotes more information than noise, while a ratio equal to infinity indicates the absence of noisy components. In this work, an independent, identically distributed Gaussian noise, yielding a SNR = 10 dB is adopted to corrupt both the training and testing data. An exemplary $\delta_1(t)$ displacement time history is reported in Fig. 3a, a as obtained from a FOM simulation for a sample seismic event; in Fig. 3b, reported an instance of the noise components mentioned above, while resulting noisy recordings is reported Fig. 3c. Before training the DL model, the data are preprocessed sensor-by-senor by standardizing all the data, so that the entire amount of data gathered bu the same sensor are normalized to feature zero mean and unit variance. Moreover, the hold-out method is used to split dataset $\mathbf{D}$ into training sets and test sets, respectively amounting to 90% and 10% of the data, while the validation set is obtained by taking 20% of the data in the training set.

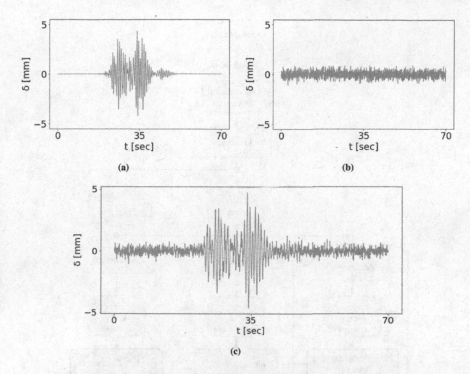

Fig. 3. Exemplary $\delta_1(t)$ displacement time histories: (a) instance of recorded signal, obtained from a FOM simulation for a sample seismic event; (b) sample of independent, identically distributed Gaussian noise affecting the sensor; (c) resulting noisy recordings. Extracted from [11].

4   Experiments and Results

The overall methodology is developed on `Google Colab` [4], a free platform based on the open-source `Jupyter` project and featuring an `NVIDIA Tesla K80 GPU` card. In Fig. 4, is illustrated a high level flowchart of the methodology. Both the adopted dataset and code are publicly released and made available at [10].

In the following, we detail the adopted CNN, which is designed to automatically and adaptively learn feature hierarchies through backpropagation, using multiple building blocks such as convolution layers, pooling layers, and fully connected layers. Then, an analysis of the performance of the DL model is presented, also considering the beneficial effect of exploiting a hyperparameter optimization strategy for DL models. Table 2 summarizes the numerosity of the target classes for the training and test sets.

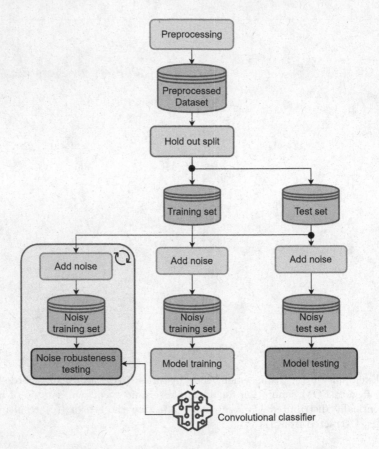

Fig. 4. High level workflow of experiments.

4.1 Damage Localization via CNN

The architecture adopted to address the damage localization task, along with its relevant hyperparameters are summarised in Table 3; specifically, it is a convolutional model made of 4 blocks. The first three deal with feature extraction, whereas the last one performs the classification task. Each of the first three blocks consists of a 1D convolutional layer, a 1D max pooling layer, and a dropout layer. The output features are then reshaped through a flatten layer and run through the classifier block, which is composed of two dense layers and a dropout one.

Adopting the Xavier's weight initialization, the loss function is minimized using the Adam algorithm, a first-order stochastic gradient descent optimizer, for a maximum of 200 allowed epochs. To ensure that the CNN does not learn the training dataset, but a possible model behind it, an early-stopping strategy is used to interrupt the learning process, whenever overfitting shows up. Whenever the loss function value computed on the validation set does not decrease for 10 epochs in a row, this latter terminates the CNN training before reaching the number of allowed epochs.

The evolution of the loss function and of the accuracy metric over the training and validation sets, obtained while training the classifier, is reported in Fig. 5a

Table 2. Classes numerosity in the training and test sets.

Damage class	Training set	Test set
Δ_0	994	117
Δ_1	1003	108
Δ_2	1008	103
Δ_3	998	113
Δ_4	992	119
Δ_5	1001	110
Δ_6	998	113
Δ_7	1005	106
Δ_8	1001	110

Table 3. Adopted CNN architecture and selected hyperparameters.

Layer type	Hyperparameter	Output shape	# parameters
Input	–	[1751, 6]	0
Conv-1D	kernels $= 6$, kernel size $= 32$	[1751, 6]	1158
MaxPool-1D	pooling size $= 8$	[219, 6]	0
Dropout	rate $= 0.15$	[219, 6]	0
Conv-1D	kernels $= 32$, kernel size $= 20$	[219, 32]	3872
MaxPool-1D	pooling size $= 6$	[37, 16]	0
Dropout	rate $= 0.15$	[37, 16]	0
Conv-1D	kernels $= 16$, kernel size $= 12$	[37, 16]	6160
MaxPool-1D	pooling size $= 4$	[10, 16]	0
Dropout	rate $= 0.15$	[10, 16]	0
Flatten	–	[160]	0
Dense	units $= 64$	[64]	10304
Dropout	rate $= 0.15$	[10, 16]	0
Dense (output)	units $= 9$	[9]	585

and Fig. 5b, respectively. The training process ends after 121 epochs due to the early stopping condition, after which the value of the tunable parameters yielding the best perfomance in terms of loss function are restored. From both graphs, it is clear that most of the gains deriving from tuning the CNN parameters are attained during the first portion of the training. The slightly irregular trend is due to the presence of dropout layers and to the stochastic nature of the minimization algorithm, for which different values of loss and classification accuracy are obtained on different mini-batches. After completing the training, the CNN achieves a global accuracy of 83%. The relevant results are gathered in the confusion matrix of Fig. 6, and in Table 4, which reports the precision and recall values class by class. We can observe that the lowest values of the assessment metrics are obtained for labels Δ_1, Δ_2 and Δ_3, suggesting that in some scenarios it may not be easy to distinguish the presence of damage on the ground floor with the undamaged state of the structure. Regarding the upper floor damage, experimental results indicate that the convolutional architecture is able to localize it with a high accuracy level.

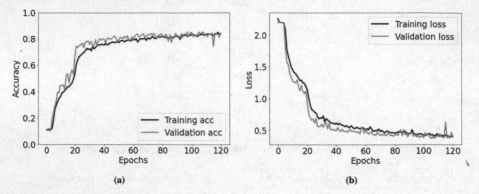

Fig. 5. Classifier training: loss function (a) and accuracy metric (b) evolution on the training and validation sets. Extracted from [11].

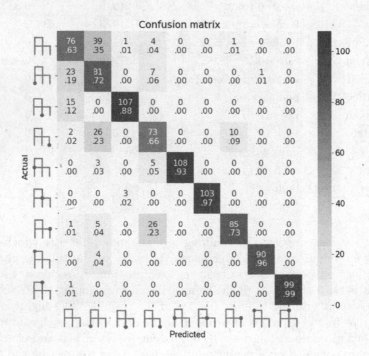

Fig. 6. Confusion matrix on test set. Extracted from [11].

4.2 Noise Tolerance Evaluation

Since the amplitude of ambient-induced vibrations is often not known a priori as it can be due to a multitude of causes, in the following the performance of the CNN model are assessed considering multiple test sets featuring a different noise level; this is useful to provide some insights on the tolerance of the trained model against noise. Overall, the model is evaluated on 13 different test sets, each collecting recordings that feature a different SNR value between 1 dB and 25 dB. The result of this analysis is reported in Fig. 7, which shows the performance of the CNN in terms of classification accuracy while varying the value of SNR. From this latter, it can be observed that the convolutional model is still able to detect damage-sensitive patterns, even when the recordings are contaminated with a very high level of noise. Specifically, when considering test sets featuring a SNR value higher than that characterizing the training data (equal to 10 dB), an improvement in the CNN performance is observed; on the other hand, when values of SNR lower than 10 DB are adopted, the CNN performance gets worse, however the attained classification accuracy attests on values largely above that one corresponding to the random guess, and still fairly remarkable for SNR values above than 5 dB.

Table 4. Damage localization Precision and Recall by class on test set.

Damage class	Precision	Recall
Δ_0	.56	.59
Δ_1	.54	.61
Δ_2	.95	.88
Δ_3	.61	.71
Δ_4	.94	.92
Δ_5	.99	.97
Δ_6	.91	.78
Δ_7	1.0	.98
Δ_8	1.0	1.0

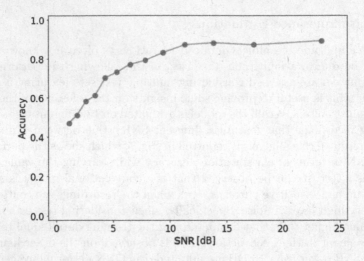

Fig. 7. Model accuracy on test set varying the EC values. Extracted from [11].

4.3   Random Search Algorithm for Hyperparameter Optimization

The design of deep neural network architectures requires to choose several parameters that are not learned during the training process but need to be selected by the user. These are the so called hyperparameters, which includes the network topology, the width/depth of each layer and the training options controlling the optimization algorithm. The performance of DL architectures critically depends on the specific choice of hyperparameters and, often, finding an optimal combination of them can make the difference between good and bad models. Therefore, to improve the performance of a DL model, a hyperparameter tuning can be carried out to find a set of optimal hyperparameters, for instance see [3]. When such a hyperparameter tuning is carried out via random search, several feasible hyperparameters configurations are iteratively instantiated to train the corresponding DL model, then the results obtained by means of each trained model are compared to determine the best set of parameters [23].

Since the DL model described in the previous section and detailed in Table 3 is defined through a simple trial-and-error heuristic, a random search optimization algorithm is here adopted to fine tune its relevant hyperprameters.

Table 5. Hyperparameter optimization: original and optimal values.

Name	Old value	Optimal value	Suitable range
Batch	128	60	[60; 128]
Drop	.15	.165	[.11; .17]
Kern-1	32	32	[25;27], [31;32], [40;48]
Kern-2	20	16	[15; 30]
Kern-3	16	5	[25; 32], [4; 13]
Pool-1	8	11	[7; 18]
POol-2	6	5	[5; 11]
Pool-3	4	3	[2; 8]
fc	64	92	[32; 128]
Acc	.83	.85	

Figure 8a shows the iterations of the random search optimization algorithm through a parallel coordinates chart; at each iteration the relevant hyperparameters are sampled in the corresponding ranges reported in the same figure, with each parameter described by a uniform probability distribution. Overall, we obtain accuracy values between 76% and 85%. The hyperparameter ranges yielding the best results are highlighted in Fig. 8b and also summarized in Table 5 along with their optimal value. Adopting the optimal hyperparameters the classification accuracy over the considered damage localization task increases from the previous value of 83% up to 85%.

Figure 9 shows the confusion matrix computed after the grid search hyperparameter optimization. We can observe the accuracy value improved for some classes in comparison to the non-optimized model. Specifically, the ground and 1st-floor classes demonstrated an improvement in accuracy, whereas 2nd-floor classes showed marginal or no change. Such classes count the highest number of misclassifications compared to the respective classes shown in the confusion matrix shown in Fig. 6.

Despite the partial improvement across classes, the optimization process improved the overall performance of the model, indicating the importance of hyperparameter tuning in machine learning algorithms.

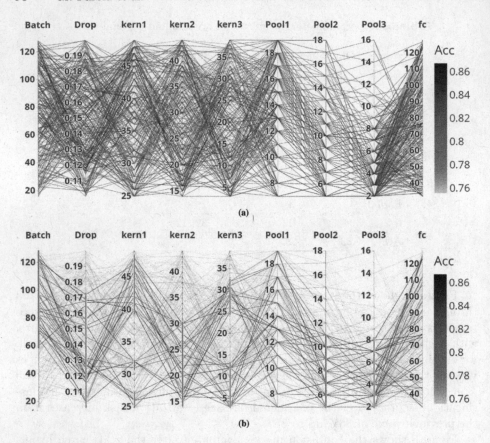

Fig. 8. Hyperparameter optimization:(a) iterations of the random search optimization algorithm; (b) identified value ranges yielding an improved performance.

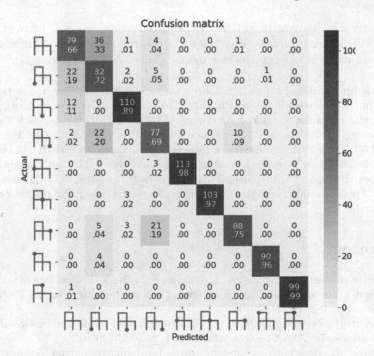

Fig. 9. Confusion matrix on test set after hyperparameter optimization.

5   Conclusion

Within a structural health monitoring framework, this work has proposed a comprehensive methodology for structural damage localization based on convolutional neural networks and a digital twins. The digital twin paradigm has been introduced to meet the need of labeled data to train deep learning models while adopting a supervised learning approach. The digital twin design has been carried out through a physics-based numerical model useful to represent the physical asset in a virtual space. A reduced-order modeling strategy is also adopted to speed up the entire procedure and allow to move toward real-time applications. Finally, a classifier based on a convolutional neural network has been adopted to perform automatic feature extraction and to relate raw sensor data to the corresponding structural health conditions. The latter has been fine tuned using the random search hyperparameter optimization algorithm.

The proposed strategy has been assessed on the monitoring of a two-story portal frame subjected to the action of seismic events. The damage localization task has been carried out with a remarkable accuracy and the method has shown to be insensitive to ambient-induced and measurement noise and to the varying operational conditions, characterized by seismic events of different nature.

This work represents a preliminary effort to demonstrate the capabilities of the proposed strategy. Beside the need of a further validation within a suitable experimental setting, the next studies will also take into account the eventuality of buildings simultaneously suffering multiple damaged zones. Moreover, a comparison with alternative deep learning architectures, such as the long-short term memory model and the transformer model, should be envisaged.

Acknowledgements. This work has been partially supported by: (i) the University of Pisa, in the framework of the PRA_2022_101 project "Decision Support Systems for territorial networks for managing ecosystem services"; (ii) the Tuscany Region, in the framework of the"SecureB2C" project, POR FESR 2014–2020, Law Decree 7429 31.05.2017; (iii) the Italian Ministry of Education and Research (MIUR), in the framework of the FoReLab project (Departments of Excellence). The authors are grateful to the research team at Politecnico di Milano composed of Alberto Corigliano, Andrea Manzoni, Luca Rosafalco and Stefano Mariani, for several insightful discussions about this research.

References

1. Aparicio, J., Jiménez, A., Ureña, J., Alvarez, F.J.: Realistic modeling of underwater ambient noise and its influence on spread-spectrum signals. In: OCEANS 2015-Genova, pp. 1–6. IEEE (2015)
2. Aydemir, H., Zengin, U., Durak, U.: The digital twin paradigm for aircraft review and outlook. In: AIAA Scitech 2020 Forum, p. 0553 (2020)
3. Bergstra, J., Bengio, Y.: Random search for hyper-parameter optimization. J. Mach. Learn. Res. **13**(2) (2012)
4. Bisong, E.: Building Machine Learning and Deep Learning Models on Google Cloud Platform. Springer, Cham (2019). https://doi.org/10.1007/978-1-4842-4470-8
5. Cimino., M., Galatolo., F., Parola., M., Perilli., N., Squeglia., N.: Deep learning of structural changes in historical buildings: the case study of the Pisa tower. In: Proceedings of the 14th International Joint Conference on Computational Intelligence, INSTICC, pp. 396–403. SciTePress (2022)
6. Galatolo, F.A., Cimino, M.G.C.A., Vaglini, G.: Using Stigmergy to incorporate the time into artificial neural networks. In: Groza, A., Prasath, R. (eds.) MIKE 2018. LNCS (LNAI), vol. 11308, pp. 248–258. Springer, Cham (2018). https://doi.org/10.1007/978-3-030-05918-7_22
7. Jones, D., Snider, C., Nassehi, A., Yon, J., Hicks, B.: Characterising the digital twin: a systematic literature review. CIRP J. Manuf. Sci. Technol. **29**, 36–52 (2020)
8. Li, D., Zhang, J., Zhang, Q., Wei, X.: Classification of ECG signals based on 1D convolution neural network. In: 2017 IEEE 19th International Conference on e-Health Networking, Applications and Services (Healthcom), pp. 1–6. IEEE (2017)
9. Paolucci, R., Gatti, F., Infantino, M., Smerzini, C., Özcebe, A.G., Stupazzini, M.: Broadband ground motions from 3D physics-based numerical simulations using artificial neural networksbroadband ground motions from 3D PBSS using ANNs. Bull. Seismol. Soc. Am. **108**(3A), 1272–1286 (2018)
10. Parola, M.: Damage localization task source code and data. https://github.com/topics/structural-health-monitoring
11. Parola., M., Galatolo., F., Torzoni., M., Cimino., M., Vaglini., G.: Structural damage localization via deep learning and IoT enabled digital twin. In: Proceedings of the 3rd International Conference on Deep Learning Theory and Applications - DeLTA, INSTICC, pp. 199–206. SciTePress (2022). https://doi.org/10.5220/0011320600003277

12. Quarteroni, A., Manzoni, A., Negri, F.: Reduced Basis Methods for Partial Differential Equations: An Introduction, vol. 92. Springer, Cham (2015)
13. Rosafalco, L., Torzoni, M., Manzoni, A., Mariani, S., Corigliano, A.: Online structural health monitoring by model order reduction and deep learning algorithms. Comput. Struct. **255**, 106604 (2021)
14. Sabetta, F., Pugliese, A.: Estimation of response spectra and simulation of non-stationary earthquake ground motions. Bull. Seismol. Soc. Am. **86**(2), 337–352 (1996)
15. Toh, G., Park, J.: Review of vibration-based structural health monitoring using deep learning. Appl. Sci. **10**(5), 1680 (2020)
16. Torzoni, M., Manzoni, A., Mariani, S.: Structural health monitoring of civil structures: a diagnostic framework powered by deep metric learning. Comput. Struct. **271**, 106858 (2022). https://doi.org/10.1016/j.compstruc.2022.106858
17. Torzoni, M., Manzoni, A., Mariani, S.: A deep neural network, multi-fidelity surrogate model approach for Bayesian model updating in SHM. In: Rizzo, P., Milazzo, A. (eds.) European Workshop on Structural Health Monitoring. EWSHM 2022. Lecture Notes in Civil Engineering, vol. 254, pp. 1076–1086. Springer, Cham (2023). https://doi.org/10.1007/978-3-031-07258-1_108
18. Torzoni, M., Rosafalco, L., Manzoni, A.: A combined model-order reduction and deep learning approach for structural health monitoring under varying operational and environmental conditions. Eng. Proc. **2**(1), 94 (2020)
19. Torzoni, M., Rosafalco, L., Manzoni, A., Mariani, S., Corigliano, A.: SHM under varying environmental conditions: an approach based on model order reduction and deep learning. Comput. Struct. **266**, 106790 (2022). https://doi.org/10.1016/j.compstruc.2022.106790
20. Wang, X., et al.: Probabilistic machine learning and Bayesian inference for vibration-based structural damage identification (2022)
21. Ye, X., Jin, T., Yun, C.: A review on deep learning-based structural health monitoring of civil infrastructures. Smart Struct. Syst. **24**(5), 567–585 (2019)
22. Yiğitler, H., Badihi, B., Jäntti, R.: Overview of time synchronization for IoT deployments: clock discipline algorithms and protocols. Sensors **20**(20), 5928 (2020)
23. Yu, T., Zhu, H.: Hyper-parameter optimization: a review of algorithms and applications. arXiv preprint: arXiv:2003.05689 (2020)

Evaluating and Improving RoSELS for Road Surface Extraction from 3D Automotive LiDAR Point Cloud Sequences

Dhvani Katkoria and Jaya Sreevalsan-Nair[✉][iD]

Graphics-Visualization-Computing Lab, IIIT Bangalore, 26/C, Electronics City, Karnataka 560100, India
jnair@iiitb.ac.in
http://www.iiitb.ac.in/gvcl

Abstract. Navigable space determination is a difficult problem encountered in robotics and intelligent vehicle technology and it requires integrated solutions using advances in computer vision and computational geometry. The conventional data processing workflow uses semantic segmentation to identify road points from three-dimensional (3D) automotive LiDAR point clouds, which have to be extended to determine its boundary points. The boundary points are critical in getting an initial coarse extraction of its surface. Hence, to filter the curb or edge points from the road points, the semantic segmentation results are postprocessed. The road surface is determined by the entire trajectory information, which is captured as a sequence of LiDAR point clouds. For the entire road surface extraction, we use an automated system, RoSELS (Road Surface Extraction for LiDAR point cloud Sequence) that works at different scales, *i.e.*, at the scale of a point, a point cloud, and the sequence of point clouds. In this paper, we evaluate the algorithms used in RoSELS, namely, the Gaussian Mixture Model for curb detection and ResNet-50 for transfer learning in frame classification. We evaluate the quality of the mesh extracted using RoSELS, which is intended for straight-road geometry. Here, we also show how the initial extracted surface can be further post-processed to extend RoSELS for curved roads.

Keywords: Road surface extraction · 3D LiDAR point clouds · Automotive LiDAR · Vehicle LiDAR · Ego-vehicle · Semantic segmentation · Ground filtering · Frame classification · Road geometry · Sequence data · Point set smoothing · Range view · Multiscale feature extraction · Local features · Mesh quality · Transfer learning · Clustering · Delaunay tetrahedralization

1 Introduction

Robotics and intelligent vehicle technology frequently encounter the need for navigable space determination, which calls for integrated solutions using

Supported by MINRO and IIIT Bangalore.

methods in computer vision and computational geometry. In 3D automotive LiDAR point clouds, the surface on which a vehicle can travel is referred to as *navigable space*, that includes the road surface. The semantic class of "ground" points usually includes various fine-grained classes, such as "road," "parking," "sidewalk," "terrain," etc. [28]. One of the motivation behind the semantic segmentation of the automotive LiDAR point clouds is its requirement for ground plane estimation, that is an initial step in road surface extraction [28,29].

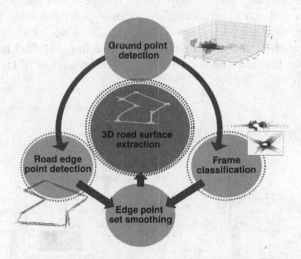

Fig. 1. Summary of RoSELS [21] for 3D road surface extraction from an automotive LiDAR point cloud sequence. Here, we rigorously evaluate algorithms used in the steps (marked by red dotted circles), and perform post-processing of the surface (marked by a blue dotted circle) to extend the application of RoSELS from straight roads to curved ones. (This figure has been extracted from our previous work [21] and modified for the current one.) (Color figure online)

Even for a single point cloud scan, the ground plane estimation is implemented discretized and piecewise. Such a surface serves as a coarse approximation of the surface geometry. However, the estimation necessitates a systematic geometric analysis, which is currently understudied [21], to compute the number, position, and orientation of the planes required to form a watertight surface. At the same time, a coarse surface mesh can be extracted using the road boundary points. This is computationally cheaper than creating a fine mesh with all road points using extensive geometric processing. However, such fine meshes are required for nuanced applications such as pothole identification for road maintenance. For a majority of applications, coarse approximation is sufficient which is uses samples of road boundary [21]. The road surface has to be extracted for a sufficient number of LiDAR scans on a trajectory to capture the entire path. Thus, an automated system for road surface extraction using a sequence of LiDAR point clouds, RoSELS [21] fills the gap in navigable space determination.

RoSELS uses a 5-step algorithm to extract the road surface using the positions of the ego-motion of the vehicle, where several methods from computer vision and computational geometry are used. The extracted road surface extraction is that which is visible from the point-of-view of the vehicle. Hence, the vehicle is called an *ego-vehicle* [29]. The steps of the algorithm are ground point detection, road edge detection, frame classification, edge point set smoothing, and 3D road surface extraction (Fig. 1). In this work, we evaluate the key methods used in these algorithms and the system itself. For the latter, we use mesh quality metrics to evaluate the system output, *i.e.,* the watertight road surface. RoSELS is designed to extract segments of straight road geometry. To extend RoSELS for curved road extraction, we propose postprocessing the surface to fill the missing road geometry, specifically as the road turns. Figure 1 shows our current contributions in evaluating and extending RoSELS in its overview, and Fig. 2 shows them in the specific steps in the workflow of the systembreak implementation.

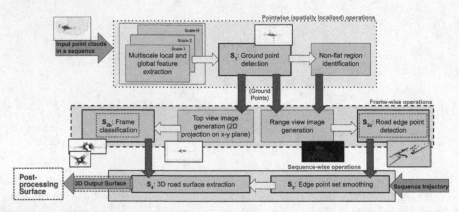

Fig. 2. The workflow of RoSELS system for road surface extraction [21], for generating 3D road surface, from input 3D LiDAR frame-wise point clouds in a sequence and its trajectory information (position and pose of the vehicle). In this work, we evaluate the intermediate steps (red dotted boxes) and add a preprocessing step (blue dotted box). (This figure has been extracted from our previous work [21] and modified for the current one.) (Color figure online)

One of the key methods in RoSELS implementation is the expectation-maximization algorithm [14] for segmenting the ground points in each point cloud to obtain curb or edge points using local height differences. Curb point detection is performed as a postprocessing step after ground point filtering [5]. Local height differences are significant features for ground point filtering [2]. The use of EM has been justified based on its use in hierarchical binary clustering for semantic segmentation of airborne LiDAR point clouds [23]. We evaluate the efficacy of EM algorithm here.

Another key method in RoSELS implementation is transfer learning using ResNet-50 [18] for frame classification. ResNet-50 is a region-based convolutional neural network (R-CNN) architecture [18], which is used for classifying natural images. Each frame, *i.e.,* each point cloud, is represented as a top-view projection image, and these images are classified based on the road geometry of the point cloud based on the ego position. We consider a coarse classification of road geometry as "straight" and "curved" roads. The images used for training and testing the R-CNN model are the visualizations of the remission values of the ground points filtered in RoSELS. Different variant of the ResNet architecture is the best performing pre-trained CNN architecture for the classification of high-resolution aerial images [24], which is similar to top-view classification. We evaluate the performance of ResNet architecture for frame classification here.

In the presence of turnings and complex topology, such as junctions, extracting road geometry becomes difficult. RoSELS extracts road surface exclusively from the road edge points. This restricts the applicability of RoSELS to straight roads, and for most sequences with a significant number of contiguous straight highways, this technique works. In our work, we first determine the quality of the mesh of the straight road segments. We then extend the system implementation to curved roads by using a postprocessing step using geometry reconstruction.

Our proposed contributions in evaluating and improving RoSELS [21] are:

1. Efficacy determination of the use of the expectation-maximization algorithm in the road edge point detection (S_{2a}).
2. Performance evaluation of the use of the ResNet-50 architecture in the frame classification step (S_{2b}),
3. System evaluation of RoSELS based on the surface mesh quality of the straight road segments in the trajectory.
4. Surface extraction for curved roads by postprocessing the road surface of the straight road segments.

2   Related Work

We evaluate the different steps in the workflow of RoSELS in road surface estimation, namely, road edge detection, scene classification based on the environment scanned by the ego-vehicle, surface estimation, and finally, surface refinement. We list the feature extraction methods, such as manual feature extraction based on analysis and feature extraction using deep learning networks, both of which are useful in tasks such as road surface extraction. We also look at the state-of-the-art in scene classification, and ground plane estimation.

2.1   Road Edge Detection

For mobile laser scanning (MLS)/LiDAR point clouds [35,42], monocular pictures captured by moving vehicles [34], and elevation maps from 2D laser scanners [25], curb extraction has been examined. To find potential points or locations, all of these techniques employ elevation filtering and the proper line-fitting algorithms. Our method is most similar to road boundary extraction for MLS point clouds [35], which is extracted by searching outward from the vehicle trajectory to discover edge points. The MLS point clouds differ, though, in that the search is carried out in the candidate point set, and we take advantage of the range image view of the vehicle LiDAR point cloud being queried.

The curb points are designated as "sidewalk" in the benchmark SemanticKITTI dataset for vehicle LiDAR point clouds, where the points are first labeled in tiles by human annotator [6] and the road boundary/curb points are subsequently specifically refined [5]. The IoU (intersection over union) score for the sidewalk is 75.5% in SCSSnet [29], and 75.2% with RangeNet++ [27] in the baseline techniques in the benchmark test of semantic segmentation performed using deep learning architectures.

Thus, while deep learning algorithms for semantic segmentation are good at identifying road points, they fall short in detecting road edge points. The class disparity can be used to explain this. Therefore, RoSELS uses a ground-based road edge recognition method that considers structure. RoSELS uses height-based hand-crafted features in supervised learning techniques [2] to determine if a road point is its edge.

2.2   Scene Classification

We look at the state-of-the-art in the area of scene categorization, which is closest to our novel frame classification. Transfer learning has been used for coarse scene classification on satellite or aerial images [44], where ResNet (Residual Network) has shown close to accurate performance [18]. ResNet with 50 layers (ResNet-50) offers the best performance and value for classifying the land cover in remote sensing images [31]. The KITTI vision benchmark suite has been used to classify the types of roads based on their functionality as "highways" and "non-highways" with the help of AlexNet [22]. RoSELS uses the ResNet-50 architecture for frame classification rather than AlexNet since it performed better on aerial images. The choice fits the need for a deep learning architecture that performs well for the top-view of the road. The top-views are such that they can be used effectively to perceptually determine the road geometry classes, *i.e.*, "straight" and "curved."

2.3   Ground Plane Estimation

According to recent research on road extraction, the geometry extraction process is preceded by ground plane estimation and segmentation [28,29]. To estimate ground elevation, GndNet creates a pseudo-image using 2D voxelization or pillars, which is then sent to a convolutional encoder-decoder [28]. SCSSnet conducts a precise ground plane estimate and employs semantic segmentation to locate ground points [29]. Since the direct extraction of coarse geometry is what we're after, we locate edge points and triangulate them with trajectory points. The *mesh map*, which is a triangulation of an automotive LiDAR using surface normals produced from range images [12], is also different from the 3D surface extraction method used in RoSELS.

2.4   Surface Reconstruction

Extracted surfaces are improved using geometric algorithms, of which a vast majority is in hole filling [17]. While our requirement here is to add a patch between meshes of straight road segments, some of the hole filling algorithms, such as modified MeshFix [17], are applicable. The original MeshFix [3] uses a single combinatorial manifold for surface reconstruction, whereas the modified algorithm fixes the errors in topology by learning how neighboring connected components relate to each other. Such methods are applicable to complex mesh geometry. Given our requirement for planar surfaces, we use the method of fitting local patches [7], where a surface patch for turning roads can be used.

Since the *medial axis* or *medial skeleton* [39] of the road surface has to match with the trajectory as given in the LiDAR point cloud sequence, the surface reconstruction is constrained. Medial axis transform using extrusion has been used in such cases [7], where *ribbons* are constructed [9].

3   Background: RoSELS System Description

RoSELS has a one-of-a-kind workflow to extract approximate road geometry, for straight roads. The workflow of road surface extraction consists of five key steps, namely, $(\mathbf{S_1})$ ground point detection, $(\mathbf{S_{2a}})$ frame classification implicitly giving the road geometry, $(\mathbf{S_{2b}})$ road edge point detection, $(\mathbf{S_3})$ edge point set smoothing, and $(\mathbf{S_4})$ 3D road surface extraction. As shown in Figs. 1 and 2:

– The point-wise operation $\mathbf{S_1}$,
– The frame-wise operations, $\mathbf{S_{2a}}$ and $\mathbf{S_{2b}}$, to be implemented in parallel, and
– The sequence-wise operations, $\mathbf{S_3}$ and $\mathbf{S_4}$, that require the trajectory information of the ego-vehicle.

We redirect the readers to our previous work [21] for the detailed system description including the pseudocode of the entire workflow.

3.1   Ground Point Detection ($\mathbf{S_1}$)

For classifying points into "ground" and "non-ground" classes, $\mathbf{S_1}$ involves two sequential sub-steps, namely, outlier removal and semantic segmentation. Here, we exploit the temporal and spatial locality of the points.

Outlier Removal. The Iterative Closest Point (ICP) registration [8] is used to do point cloud registration or scan matching on two separate point clouds to determine the correspondence pairs of points between the two. To exploit temporal locality, the point correspondence pairings in successive frames are determined. The points that do not get registered are marked as "outliers" and are filtered out. Iterative registration is used on three successive frames at a time due to the consistency of motion across frames in two phases. For frame (x), the registration between frames $(x-1)$ and $(x-2)$ is used to filter outliers in the frame $(x-1)$ in the first phase. In the second phase, registration between (x) and the filtered $(x-1)$ is similarly used to remove outliers in the frame (x).

Semantic Segmentation. The height-based features derived from the elevation (z) of scene objects are optimal for distinguishing ground locations from others [2]. The local and global spatial features, which are handcrafted, are extracted first. These features are fed into a supervised learning model; namely, the Random Forest Classifier (RFC) [10] to classify as ground or otherwise.

The multi-scale local features are used to improve the accuracy of classification using supervised learning [40]. The local neighborhood with a hybrid search that combines the criteria of the spherical and the k-nearest neighborhoods (knn) is uniformly queried for all points. Then, local and global height features used for ground point detection are the same as listed in Table 1. These extracted features are computed and used in an RFC for both training and testing.

3.2   Road Edge Point Detection (S_{2a})

The ground points that physically interface with the curb or sidewalk are those spots along the borders of the road [5]. Both the left and right banks of the road must be extracted for road surface extraction. Edge detection using gradient information is widely used in image processing and computer vision, where it is used to identify image borders. The height gradient is a statistically significant feature to identify the ground points [2], which is repurposed for finding road boundaries, as used in image processing.

Table 1. Point-wise features at each scale for ground detection (S_1) as used in RoSELS [21].

Local Features	Global Features	Global Features
Point-based		Frame-based
Height-based – Difference from max. – Difference from mean – Standard deviation	*Height-based* – Value (z-coordinate) – Difference from mean (*of frame*) *Position-based* – Distance from sensor – Elevation angle θ	*Height-based* – Difference from mean – Standard deviation

Height Gradient-Based Differentiation. Here, only the ground points are used to compute the first-order derivatives or gradients of height values. The height gradients are determined using height difference features computed in the local neighborhood. RoSELS implements binary clustering of points as the ground point either belong to "flat" and "non-flat" surfaces. These surfaces have low and high height gradients, respectively. From the exhaustive list of hand-crafted features needed to semantically segment 3D aerial and terrestrial LiDAR point clouds [40], RoSELS uses two appropriate height-difference (Δ_z) features computed in a local neighborhood. The first feature is computed directly from the neighborhood in 3D space, and the second one indirectly by using the local neighborhood in the 2D accumulation map computed at each point. The 2D accumulation map generates local neighborhoods of points projected to xy-plane, within a square of fixed length (*e.g.*, 0.25 m), centered at the point. For the clustering process, we observe that the clear clusters do not exist in automotive LiDAR point clouds. We note that the vehicle LiDAR point clouds do not have distinct clusters for the clustering process. RoSELS uses the Expectation-Maximization algorithm [14] for binary clustering.

Projection to Range Images. The edge points are in the non-flat zones, and the road edge points preferred in RoSELS are those that are closest to the centerline or the trajectory. The frame-wise range image is the ideal projection usable in a frame-by-frame process of centerline detection. RoSELS uses the column of the range image where the sensor, or the ego-vehicle, is located as the centerline. The range image is the dense rasterized representation of the obscured view from the ego-vehicle. Hence, it is constructed as a spherical projection of the points closest to the ego-vehicle, with the pixels colored according to a chosen attribute of the closest point in the pixel. The angular resolution in the elevation and azimuthal angles determines the image resolution. For instance, the angular resolution of the Velodyne HDL-64E S2 used in SemanticKITTI data [6] acquisition is given by 64 angular subdivisions (*i.e.,* $\approx 0.4°$) in the elevation angle spanning for 26.8°, and similarly 0.08° angular resolution for 360° azimuthal angle. Thus, the range image has a 64×4500 resolution.

Edge Detection. RoSELS uses a *scanline algorithm* on the range image in order to determine the edge points. While scanning the image of size $H \times W$ row-wise, the key positions in the pixel space, relative to the ego-vehicle, are:

- at P_{cf}, *i.e.,* $(0, \frac{W}{2})$, which indicates the centerline column in the front;
- at P_{cbL}, *i.e.,* $(0, 0)$, which indicates the centerline column in the back (rear), but on the left-side of the ego-vehicle; and
- at P_{cbR}, *i.e.,* $(0, W)$, which indicates the centerline column in the back, but on the right-side.

The left and right sides of the ego-vehicle are considered with respect to its front face. In RoSELS, the pixels in each row are scanned in the appropriate direction until a pixel carrying a non-flat region point is encountered. The pixels on the centerline column(s) are utilized as the reference pixel for each row. For the front left and right pixels representing non-flat region points, the scanline traverses from P_{cf} towards P_{cbL} and P_{cbR}, respectively. Similarly, on the rear side, for the left side, it traverses from P_{cbL} to P_{cf}, and for the right side, from P_{cbR} to P_{cf}. After locating these pixels, their corresponding 3D LiDAR points are to be determined. These row-wise points are referred to as p_{fL}, p_{bL} on the left side, and p_{fR}, p_{bR} on the right side. These frame-wise points are added to the side-specific sets, EP_L and EP_R, for the left and right sides, respectively.

3.3   Frame Classification (S2b)

The surface extraction technique is influenced by the underlying road shape as expected. The road edge points are used in RoSELS to create triangulated (surface) meshes. The points along the road boundary must be sampled sufficiently for precise edge extraction. The curvature of the road influences this sampling.

The first level of road geometry classification includes the "straight" and "curved" roads as shown in Fig. 3(a). For three reasons, RoSELS is limited to extracting the geometry of straight roads. Firstly, curved roads require more samples as edge points in order for the edges to be retrieved accurately, and the sample quantity is decided using geometric techniques. Secondly, from the perspective of the ego-vehicle, the wider road topology is not sufficiently recorded to extract the curved road edges piece-wise. The existing procedure is unable to capture the road topology for turnings and crossroads, which involves T- and X-intersections that must be recorded. Thirdly, more inner road points are required in order to correctly extract curved road surfaces.

Transfer Learning Using Pre-trained Architectures. It has been observed that each frame in the 3D LiDAR point cloud sequence clearly illustrates the road geometry from its top-view, *i.e.,* a 2D projection on the x-y plane. RoSELS uses transfer learning with ResNet-50 in order to take advantage of the perceptual variations across frames. Effective scene classification of perceptually distinguishable images has been achieved for other works using ResNet-50, which

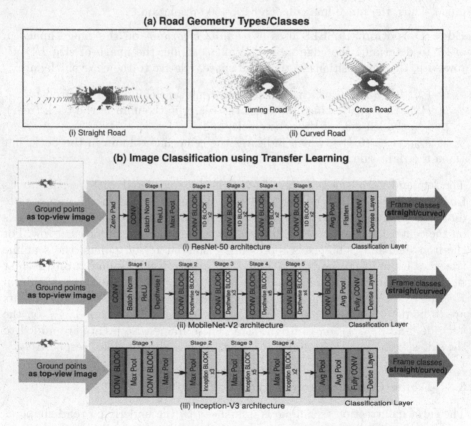

Fig. 3. (a) The road geometry classification is hierarchical with straight and curved road types. (b) Frame classification in RoSELS [21] is implemented using transfer learning on top-view images of ground points in each frame. In this paper, we add a comparative analysis of ResNet-50 with MobileNet-V2 and Inception-V3 architectures.

is a region-based convolutional neural network (R-CNN) architecture [18]. We generate the top view image of ground points for classification and classify these images based on road geometry types (Fig. 3(a)), *i.e.*, straight and curved.

The 2D top-view RGB image is rendered using the attribute values of the points using sequential colormaps. We use the grayscale map and a perceptually uniform sequential colormap, namely the Viridis colormap. The sequential colormap is discretized into a specific number of bins, *e.g.*, 5 bins. The colormap is used to render the ground points found in S_1 while taking into account their remission values. Transfer learning is implemented on these images for frame classification in RoSELS (Fig. 3(b)). Here, pre-trained weights for image classification of ImageNet [15] are used, which is the de facto standard in transfer learning on images.

3.4   Edge Point Set Smoothing (S_3)

Now, the road edge points identified in S_{2a} are fitted to form edges. Due to the noise in the edges, these edges have to be smoothed. To prevent filtering out essential locations, the smoothing is carried out individually for the left and right sides of the road. Edge labeling is the process of identifying edges and removing false positives. Thus, S_3 includes both smoothing and labeling. The global coordinate system, which contains the whole trajectory of the series, is used to implement edge processing. The transformation of the coordinate system is therefore the initial sub-step.

Local to World Coordinate System Transformation. This transformation makes sure the smoothed edge is present in the 3D world space exactly as it is. The smoothing and transformation procedures are also *non-commutative* meaning that the sequence in which they are carried out must be rigorously preserved. Hence, we now add the trajectory information as an input to the workflow (Fig. 2. This input contains the position and poses of the ego-vehicle at each frame of the sequence. Transformation matrices are used to represent the shift in position and posture. Each frame's edge points are subjected to these matrices in order to translate them into the 3D world space.

Point Set Smoothing and Labeling. Using the converted coordinates, the straight road edges are smoothed. First, we identify the subsequences of frames that make up consecutive stretches of straight roads. Each side's implementation of this is done individually. The random sample consensus (RANSAC) line fitting model [16] is applied to each such subsequence. As a result, both sides of the road have disconnected smooth line segments resembling dashed lines.

3.5   3D Road Surface Extraction (S_4)

Following smoothing, each continuous segment of the left and right edge points of straight roads is used to create a triangulated mesh that represents the road surface. Here, a constrained Delaunay tetrahedralization [33] is implemented and is followed by the extraction of the outer/external surface of the tetrahedral mesh. When opposed to using projections of the 3D points for 2D Delaunay triangulation, this produces triangles of higher quality.

4   Evaluating and Improving RoSELS

Here, we propose evaluating and extending RoSELS, as detailed below.

Evaluation of EM Algorithm in S_{2a}. Here, we compare the efficacy of a *model-based clustering* method such as Expectation-maximization with a *partitional clustering* method [1], for which k-means algorithm [26] is widely used. The family of partitional clustering of items is governed by three characteristics: (a) a center that is not necessarily an item represents each cluster, (b) a distance measure to be applied on feature vectors of all items, and (c) a cost function is used for optimization, that usually is about minimizing intra-cluster distances and maximizing inter-cluster distances. Model-based clustering methods assume that there exists a model based on data distribution known for similar applications or theoretically. They assume that the given dataset closely matches the chosen model. The model inherently includes clear clusters. The remaining families of clustering includes hierarchical, neural network-based, and others are not applicable in our case.

Evaluation of ResNet-50 Architecture in S_{2b}. We evaluate two distinct frame classification models that are matched to the various levels of the road geometry class hierarchy (Fig. 3). The first model is used to categorize roads as straight or curved, and the second is used to categorize roads as straight, crossroads, and turnings. We evaluate the performance of transfer learning using two models *i.e.,* ResNet-50 as used in RoSELS and MobileNet [20], where we use transfer learning with the weights obtained on training on ImageNet dataset [15]. We perform this experiment only on the first class hierarchy level *i.e.,* "straight" and "curved" roads (Fig. 3(a)). The deep layers used in the two models can be compared in Fig. 3(b) to visualize their differences.

Here, we also consider Inception [37] as an alternative model for transfer learning. While Inception-V3 [38] gives better classification results, it is also not as efficient as Resnet-50 [19]. ResNet-50 outperforms MobileNet in classification accuracy for aerial image dataset [41], whereas MobileNets-V2 [30] performs better than Resnet-50. Hence, we compare the performance of ResNet-50, Mobilenet-V2, and Inception-V3 in RoSELS.

Self-attention module in the ResNet architecture is known to improve the classification of remote sensed images [41], and we consider the same in the scope of future work.

Evaluation of Surface Mesh Quality. In order to evaluate the surface mesh quality, conventionally the quality metric is checked for each mesh element and the collective value of the quality metric is considered for the entire mesh. For triangular meshes, an ideal linear element, *i.e.,* a triangle, is the one that is closest to the equilateral triangle, *i.e.,* the property of being *nearly equilateral.* Focusing on interior angle-based quality metrics, it is known that the maximum angle of a triangle is a better measure of its quality than its minimum angle [32]. The maximum angle being close to $180°$ is considered more harmful than the minimum angle being close to $0°$ [4,32].

Thus, we propose the use of the maximum angle criterion for the quality of triangles in the surface mesh. We then visualize the maximum angle θ_{max} values in a mesh in a sequence and also perform descriptive statistical analysis of the θ_{max} in each mesh.

Extending RoSELS to Curved Road Extraction. The trajectory has straight and curved road segments. Hence, RoSELS gives surfaces for the straight roads. We can now use piecewise linear methods to reconstruct the gaps in the road surface. This enables the reconstruction of the road surface as navigated by the ego-vehicle. Thus, it does not construct the intersecting roads in a crossroad.

The trajectory of the ego-vehicle serves as the *medial skeleton* [39] of the road surface. Once we construct the medial axis using the trajectory information, we focus on regions which are gaps between the linear segments. We then extrude the line to generate a ribbon and overlay the ribbon on the RoSELS output, *i.e.,* the surface of the straight roads. We overlay the extruded ribbon over the straight roads, and merge them, thus generating a watertight surface.

We also experiment filling the surface gap by using 3D Delaunay triangulation (or tetrahedralization). We compare the results between the extruded surface and the Delaunay triangulated surface.

5   Experiments and Results

Sequences of LiDAR point clouds that follow the path of the vehicle must be included in the input dataset for road surface extraction, together with sufficient annotations to enable machine learning algorithms. SemanticKITTI [6] works well as our test data in this regard.

Table 2. Specifications of the SemanticKITTI [6] sequences used for training and validation/testing in road surface extraction [21]

Seq. ID	# Frames					Ground truth (GT)		
Training	Total	Used	Straight	Crossroad	Turning	# Points	# "Ground" points	"Road"* (%)
00	4,541	455	238	205	12	55,300,603	21,242,723	45.2
01	1,101	111	69	23	19	11,737,924	6,684,753	71.8
02	4,661	467	195	134	138	58,678,800	26,955,344	42.8
03	801	81	17	48	16	10,038,550	4,563,802	48.8
04	271	28	25	3	0	3,518,075	1,816,228	65.9
05	2,761	277	172	102	3	34,624,816	14,025,815	40.5
06	1,101	111	54	57	0	13,567,503	8,417,991	34.1
07	1,101	111	54	57	0	1,3466,390	5,301,837	48.1
09	1,591	160	42	39	79	19,894,193	9,313,682	45.0
10	1,201	121	64	32	25	15,366,254	5,608,339	43.7
All	**19130**	**1922**	**930**	**700**	**292**	**236,193,108**	**103,930,514**	**45.3**
Testing								
08	4071	408	261	124	23	50,006,369	21,943,921	40.3

* Our annotation of "ground" combines five classes, namely, "road," "parking," "sidewalk," "terrain," and "other ground," as given in SemanticKITTI dataset [6]. Percentage values in columns 9 and 11 give the fraction of ground points in columns 8 and 10, respectively, that are annotated as "Road" in SemanticKITTI.

5.1   Implementation of Road Surface Extraction

- The road surface extraction has been implemented on Intel core i7 CPU with 12 GB of RAM. We have used Open3D library APIs [43] for point cloud registration in S_1. For neighborhood computation in S_1 and S_3, Open3D KDTree has been used. The scikit-learn library APIs [11] have been used for implementing the RFC and GMM models in S_1 and S_{2a}, respectively. Frame classification model in S_{2b} has been implemented using Keras APIs [13] and the model has been trained for five epochs. Edge point set smoothing in S_3 has used the RANSAC model from the scikit-image library. PyVista library APIs [36] have been used for geometry computation in S_4 and for extending RoSELS in this work.

5.2   Experiments and Results

Here, we discuss the dataset used, the experiments, and their results.

Dataset. The SemanticKITTI dataset [6] has been published primarily for three benchmark tasks, namely semantic segmentation and scene completion of point clouds using single and multi-temporal scans. Since our work is different from the benchmark tasks, validation is not readily available for the dataset. Given its fit as input to road surface extraction, we use SemanticKITTI for our experiments and provide an appropriate qualitative and quantitative assessment.

The SemanticKITTI dataset comprises of the training sequence IDs, 00 to 10 (Table 2). We have used sequence 08 as the validation/test set, as prescribed by the data providers, thus training our model on the remaining training sequences for our classifier models, *i.e.*, RFC model for ground point detection (S_1), and transfer learning model for frame classification (S_{2b}). We have only used every 10th frame of training sequences of SemanticKITTI since frames are captured in 0.1s and our subsampling ensures significant variations in the vehicle environment are captured without incurring high computational costs. We have found that including more overlapping data resulted in increased computation without adding new information.

We group the classes to form the ground and non-ground classes. The curbs of the road are labeled as sidewalk [5] and are important in our evaluation. For the frame classification, all frames in all training sequences, *i.e.*, from 00 to 10, are manually annotated into "straight," "crossroad," and "turning" in RoSELS implementation [21].

Experimental Setup. For multi-scale feature extraction for ground point detection using RFC, hybrid criteria for neighborhood determination have been used for three different scales. RoSELS implementation has used the constraint of r of $1m$ for the spherical neighborhood in all the scales, and variable k values for the knn neighborhood, *i.e.*, $k = 50, 100, 200$ neighbors. The original implementation has systematically experimented with several combinations of neighborhood criteria to arrive at this parameter setting. Similarly, appropriate hybrid criteria, *i.e.*, $r = 1m$ and $k = 50$ has been used for determining the local neighborhood of ground points. These neighborhoods are used for computing height-difference features used in binary clustering algorithms (EM and k-means) for detecting flat and non-flat regions from filtered ground points. For road surface extraction, we use the training dataset sequences 01, 05, 07, and 08, as done in our previous work [21], which consists of straight road segments. Additionally, we extend RoSELS for curved roads, for which we use the training sequences 03 and 10. The details of the sequences are in Table 2.

Here, we first summarize the significant results of running RoSELS from our previous work [21] for the sake of completion, and then demonstrate the results of evaluation in our current work (Fig. 4, Table 3, and Figs. 5, 6 and 7).

Summary of Results of Step-Wise Evaluation [21]. In our previous work, we evaluated each step in the algorithm (Sect. 3) and concluded the following:

- Using multiscale local features in the random forest classifier improved the accuracy of ground point classification from 96.37% to 96.63%, and furthermore, along with the outlier removal, the accuracy improved to 96.91%, *i.e.*, corresponding to mIoU (mean Intersection over Union score) 90.79% for the sequences used here.
- Using a class hierarchy of road geometry at the first level, *i.e.*, straight and road classes, gave an accuracy of classification of 82.35% as opposed to 78.51%, when using both hierarchical levels (Fig. 3(a)).
- A large number of points identified as road edge points in RoSELS belong to the "road" class as given in the annotated data in SemanticKITTI. For some sequences, a large part of the road edge points belongs to the "sidewalk" class. Thus, the road edge points identified by RoSELS are valid points.

Results of Evaluating EM Algorithm for Binary Clustering. Our results of comparing EM algorithm with k-means are qualitatively visualized in Fig. 4. In the visualizations, we observe the flat and non-flat regions identified using the k-means and EM algorithms across different road structures. Qualitatively, the visualizations show that the EM algorithm detects the boundary between regions more sufficiently and accurately than the k-means algorithm.

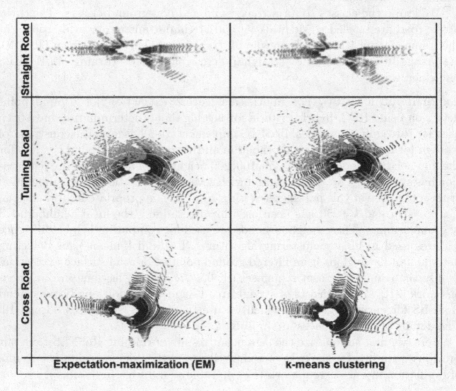

Fig. 4. Flat and non-flat regions identified for different road structures using the k-means and EM algorithms. Flat and non-flat regions are highlighted in purple and yellow respectively. (Color figure online)

The natural tendency of the clusters to form two modes in their height distribution guarantees the success of the EM algorithm as it assumes the presence of a Gaussian Mixture Model (GMM) in the data. Thus, we conclude that the flat and non-flat regions approximately form a bimodal data distribution in the 2D feature space and EM algorithm is the best suited for this requirement than an Euclidean distance-based partitional method, such as k-means.

Table 3. Frame classification results using 'Viridis' colormap for RGB image and 'binary' colormap for grayscale image with different number of bins for sequential (for numerical data) colormap and the different transfer learning models on SemanticKITTI dataset [6].

Image Type	#Bins	Resnet-50	MobileNet-V2	Inception-V3
RGB (with local maxima and minima) Using Viridis colormap	5	**88.50%**	74.50%	81.76%
	10	85.37%	71.13%	82.25%
	20	84.13%	70.13%	77.00%
	No Binning	85.25%	70.88%	79.37%
Grayscale (with local maxima and minima) Using binary colormap	5	79.37%	79.37%	79.62%
	10	76.13%	73.37%	76.25%
	20	73.25%	60.00%	71.88%
	No Binning	79.50%	78.37%	**79.63%**

Results of Evaluating ResNet-50 for Transfer Learning. For frame classification, we evaluate the performance of widely used CNN models, namely, ResNet-50, Mobilenet-V2, and Inception-V3 (Table 3). Here, we run experiments on the input data using different color schemes, namely, the grayscale images using binary colormap and the perceptually ordered Viridis colormap. We also experiment with different bins to form sequential colormap for numerical data.

The results show that ResNet-50 performs better than MobileNet-V2. When using the Viridis colormap, Inception-V3 model underperforms compared to ResNet-50 architecture. But for grayscale colormap, we observe that Inception-V3 model performs marginally better than the ResNet-50 model in all scenarios except the case of 20 bins. Given that the Inception model is not efficient for training and also do not give consistent performance, we conclude that the ResNet-50 architecture is the most optimal model for frame classification through transfer learning in RoSELS.

Results of Surface Mesh Evaluation. Our results for all the sequences are in Figs. 5 and 6. The root means square error between the edge point sets computed using "road" (GT) and "ground" (detected in S_1) points are calculated after performing these three steps on the "road" points in the ground truth (GT). In our previous work [21], we performed this analysis due to the lack of ground truth of edge points and retrieved surface. In Fig. 5 the RMSE values for Sequences 01, 05, 07, and 08 are provided. We notice that the RMSE errors are comparatively small, given that each frame has an extent of 51.2 m in front of the car and 25.6 m on either side [6].

Figure 5 shows the results of the 3D extracted surface for trajectories of various sequences. Our qualitative findings demonstrate that road surface extraction is effective on closed trajectories, complex trajectories, and straight roads with predominantly contiguous straight road segments. For the purpose of illustrating their similarities, our results of surfaces produced using detected ground points and "road" (GT) points are overlaid. We observe that the triangulated meshes have adequately covered brief turning segments and connections between straight road segments.

We also visualize the overall mesh quality based on the θ_{max} of individual triangles in Figs. 5 and 6. We observe that the visualization of element-wise θ_{max} uses the Viridis colormap. The lower end of the spectrum, *i.e.*, the blue elements are of better quality than the higher end, *i.e.*, the yellow elements. We observe that meshes generated using RoSELS have triangles of good and medium quality in Fig. 5(E). The distribution of the θ_{max} closer to the 60° can be perceived to be a better mesh, as seen in the box-and-whisker plots (Fig. 7). This behavior can be attributed by the use of Delaunay triangulation that optimizes the minimum angle θ_{min} in each element [32]. We conclude that further mesh refinement will help in improving the quality, and will be undertaken in the future study.

Results of Surface Reconstruction for Curved Roads. We shift our focus to our results for curved road surface reconstruction, as shown in Fig. 6. RoSELS after S_4 cannot extract road surfaces where (a) straight road segments are substantially fragmented and (b) subsequences contain portions of curved and straight roads that are comparable in length. Road surface extraction only extracts a portion of the road surface when edge points are not detected for significant portions of the road, as is the case in training sequences 03 and 10.

For curved roads, our surface post-processing step yields better quality mesh for extruded surfaces than the Delaunay triangulation of the surface gap, as shown in Fig. 6. This is evident from the blue This can be attributed to the higher quality of meshes when performing piecewise extraction as a first cut, where the method is aware of the transitions in the road geometry. When performing Delaunay triangulation with an optimal alpha value for all elements, we observe that the reconstruction produces *fan-like* geometry, which is not desirable. We observe also that the mesh quality is better for the method with extruded mesh than with the Delaunay triangulation, as there are more blue than yellow triangles (Fig. 6(e) and (f)). We also observe in the box-and-whisker plot (Fig. 7(b)) that the θ_{max} in meshes from the extrusion method are closer to 60° than those from the Delaunay triangulation method.

In such a scenario of curved roads, an adaptive method which is aware of the road geometry transitions may perform better when the alpha setting for Delaunay triangulation is data-driven.

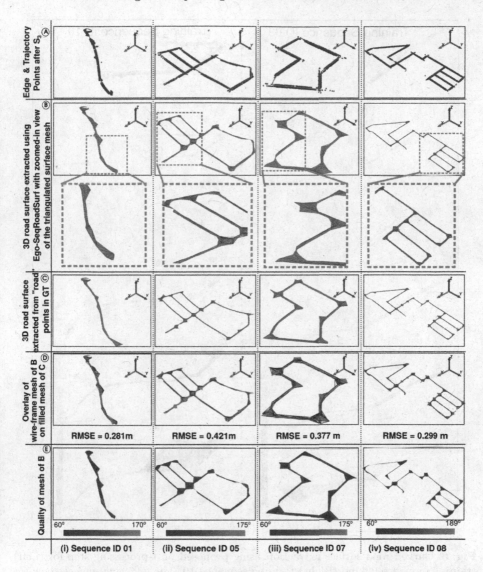

Fig. 5. Results of implementing road surface extraction [21] on training sequences of SemanticKITTI [6], where 01, 05, and 07 have been used for training our learning models, and 08 have been used for validation/testing. Row A follows the color scheme of red, purple, and green points showing the trajectory, left and right edge points, respectively. For rows B, C, and D, wireframe meshes are shown in indigo, and filled meshes are shown in tan color. Row E shows the quality of the meshes generated. (Color figure online)

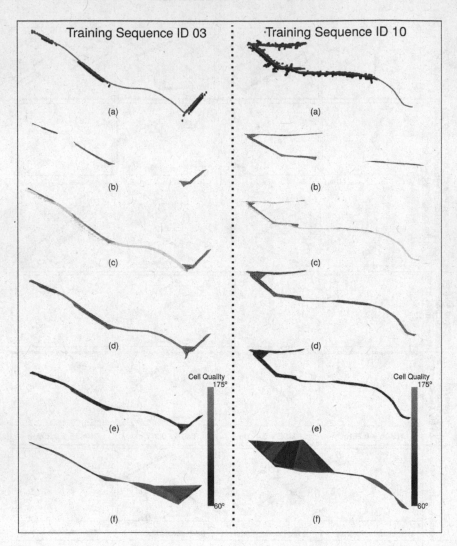

Fig. 6. Curved road surface extraction using proposed post-processing step for (Left) training sequence 03 and (Right) training sequence 10. The following images show (a) points with edge smoothing along with medial axis, (b) surface reconstructed using RoSELS, (c) extruded surface overlaid on the surface from RoSELS, (d) extruded surface merged with the RoSELS one, thus giving a piecewise linear reconstruction, (e) θ_{max} quality of mesh using extrusion, (f) θ_{max} quality of mesh using Delaunay tetrahedralization using alpha $= 150$.

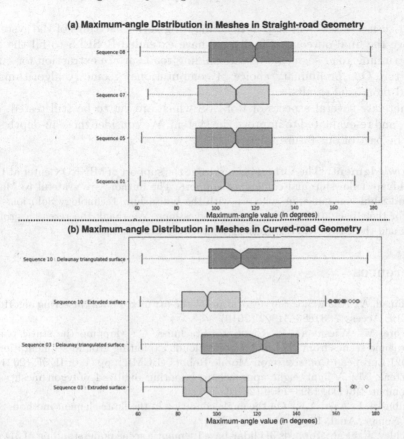

Fig. 7. Box-and-whisker plots showing the maximum angle distribution in the meshes extracted for sequences with (a) straight-road, and (b) curved-road geometries.

6   Conclusions

For a series of 3D automobile LiDAR point clouds, we implemented road surface extraction using a unique approach for automating 3D road surface extraction. The system designed and implemented, named RoSELS, is the contribution of our previous work. RoSELS has a five-step workflow that filters ground points, identifies boundary points, and extracts road surface in the form of triangular mesh for straight road segments in a trajectory. The trajectory data is in the form of a sequence of 3D LiDAR point clouds. In our previous work, the evaluation of RoSELS on four SemanticKITTI sequences with varying degrees of geometry complexity showed promising results that have been both qualitatively and quantitatively confirmed.

    In this work, we have evaluated key methods used in RoSELS and confirmed that EM algorithm and ResNet-50 architecture are optimal choices for edge point detection and frame classification, respectively. We have provided a metric to test the quality of the mesh given as an output from RoSELS. We observed that having such a metric will help in re-configuring various settings of the system to improve the final outcome. Finally, we have extended RoSELS to fill the holes in the turning road segments, thus, enabling road surface extraction for curved roads too. Our preliminary choice of computational geometry algorithms has yielded promising results.

There are several aspects of RoSELS which are yet to be still tested, optimized, and re-evaluated to improve the system. We consider those in-depth studies to be part of our future work.

Acknowledgment. The authors acknowledge the support of MINRO Center at IIITB for graduate fellowship and conference support. The authors are grateful to Mayank Sati and Sunil Karunakaran, who are with the Ignitarium Technology Solutions, Pvt. Ltd., for their insightful discussion and suggestions. We thank the research group at GVCL and the entire IIITB fraternity for their constant support.

References

1. Ahmad, A., Khan, S.S.: Survey of state-of-the-art mixed data clustering algorithms. IEEE Access **7**, 31883–31902 (2019)
2. Arora, M., Wiesmann, L., Chen, X., Stachniss, C.: Mapping the static parts of dynamic scenes from 3D LiDAR point clouds exploiting ground segmentation. In: 2021 European Conference on Mobile Robots (ECMR), pp. 1–6. IEEE (2021)
3. Attene, M.: A lightweight approach to repairing digitized polygon meshes. Vis. Comput. **26**(11), 1393–1406 (2010)
4. Babuška, I., Aziz, A.K.: On the angle condition in the finite element method. SIAM J. Numer. Anal. **13**(2), 214–226 (1976)
5. Behley, J., et al.: Towards 3D lidar-based semantic scene understanding of 3D point cloud sequences: the semantickitti dataset. Int. J. Robot. Res. **40**(8–9), 959–967 (2021)
6. Behley, J., et al.: SemanticKITTI: a dataset for semantic scene understanding of LiDAR sequences. In: Proceedings of the IEEE/CVF International Conference on Computer Vision, pp. 9297–9307 (2019)
7. Berger, M., et al.: State of the art in surface reconstruction from point clouds. In: Eurographics 2014-State of the Art Reports, vol. 1, no. 1, pp. 161–185 (2014)
8. Besl Paul, J., McKay, N.D.: A method for registration of 3-D shapes. IEEE Trans. Pattern Anal. Mach. Intell. **14**(2), 239–256 (1992)
9. Biasotti, S., et al.: Skeletal structures. In: De Floriani, L., Spagnuolo, M. (eds.) Shape Analysis and Structuring, pp. 145–183. Springer, Heidelberg (2008). https://doi.org/10.1007/978-3-540-33265-7_5
10. Breiman, L.: Random forests. Mach. Learn. **45**(1), 5–32 (2001)
11. Buitinck, L., et al.: API design for machine learning software: experiences from the scikit-learn project. In: ECML PKDD Workshop: Languages for Data Mining and Machine Learning, pp. 108–122 (2013)
12. Chen, X., Vizzo, I., Läbe, T., Behley, J., Stachniss, C.: Range image-based LiDAR localization for autonomous vehicles. In: 2021 IEEE International Conference on Robotics and Automation (ICRA), pp. 5802–5808. IEEE (2021)
13. Chollet, F., et al.: Keras (2015). https://keras.io/

14. Dempster, A.P., Laird, N.M., Rubin, D.B.: Maximum likelihood from incomplete data via the EM algorithm. J. Roy. Stat. Soc.: Ser. B (Methodol.) **39**(1), 1–22 (1977)

15. Deng, J., Dong, W., Socher, R., Li, L.J., Li, K., Fei-Fei, L.: ImageNet: a large-scale hierarchical image database. In: 2009 IEEE Conference on Computer Vision and Pattern Recognition, pp. 248–255. IEEE (2009)

16. Fischler, M.A., Bolles, R.C.: Random sample consensus: a paradigm for model fitting with applications to image analysis and automated cartography. Commun. ACM **24**(6), 381–395 (1981)

17. Guo, X., Xiao, J., Wang, Y.: A survey on algorithms of hole filling in 3D surface reconstruction. Vis. Comput. **34**(1), 93–103 (2018)

18. He, K., Zhang, X., Ren, S., Sun, J.: Deep residual learning for image recognition. In: Proceedings of the IEEE Conference on Computer Vision and Pattern Recognition, pp. 770–778 (2016)

19. He, T., Zhang, Z., Zhang, H., Zhang, Z., Xie, J., Li, M.: Bag of tricks for image classification with convolutional neural networks. In: Proceedings of the IEEE/CVF Conference on Computer Vision and Pattern Recognition, pp. 558–567 (2019)

20. Howard, A.G., et al.: MobileNets: Efficient Convolutional Neural Networks for Mobile Vision Applications. arXiv preprint arXiv:1704.04861 (2017)

21. Katkoria, D., Sreevalsan-Nair, J.: RoSELS: road surface extraction for 3D automotive LiDAR point cloud sequence. In: Proceedings of the 3rd International Conference on Deep Learning Theory and Applications (DeLTA), pp. 55–67. INSTICC, SciTePress (2022). https://doi.org/10.5220/0011301700003277

22. Krizhevsky, A., Sutskever, I., Hinton, G.E.: Imagenet classification with deep convolutional neural networks. In: Advances in Neural Information Processing Systems, vol. 25 (2012)

23. Kumari, B., Sreevalsan-Nair, J.: An interactive visual analytic tool for semantic classification of 3D urban LiDAR point cloud. In: Proceedings of the 23rd SIGSPATIAL International Conference on Advances in Geographic Information Systems, pp. 1–4 (2015)

24. Liang, Y., Monteiro, S.T., Saber, E.S.: Transfer learning for high resolution aerial image classification. In: 2016 IEEE Applied Imagery Pattern Recognition Workshop (AIPR), pp. 1–8. IEEE (2016)

25. Liu, Z., Liu, D., Chen, T., Wei, C.: Curb detection using 2D range data in a campus environment. In: 2013 Seventh International Conference on Image and Graphics, pp. 291–296. IEEE (2013)

26. Lloyd, S.: Least squares quantization in PCM. IEEE Trans. Inf. Theory **28**(2), 129–137 (1982)

27. Milioto, A., Vizzo, I., Behley, J., Stachniss, C.: Rangenet++: fast and accurate lidar semantic segmentation. In: 2019 IEEE/RSJ International Conference on Intelligent Robots and Systems (IROS), pp. 4213–4220. IEEE (2019)

28. Paigwar, A., Erkent, Ö., Sierra-Gonzalez, D., Laugier, C.: Gndnet: fast ground plane estimation and point cloud segmentation for autonomous vehicles. In: 2020 IEEE/RSJ International Conference on Intelligent Robots and Systems (IROS), pp. 2150–2156. IEEE (2020)

29. Rist, C.B., Schmidt, D., Enzweiler, M., Gavrila, D.M.: SCSSnet: learning spatially-conditioned scene segmentation on LiDAR point clouds. In: 2020 IEEE Intelligent Vehicles Symposium (IV), pp. 1086–1093. IEEE (2020)

30. Sandler, M., Howard, A., Zhu, M., Zhmoginov, A., Chen, L.C.: Mobilenet V2: inverted residuals and linear bottlenecks. In: Proceedings of the IEEE Conference on Computer Vision and Pattern Recognition, pp. 4510–4520 (2018)

31. Scott, G.J., England, M.R., Starms, W.A., Marcum, R.A., Davis, C.H.: Training deep convolutional neural networks for land-cover classification of high-resolution imagery. IEEE Geosci. Remote Sens. Lett. **14**(4), 549–553 (2017)
32. Shewchuk, J.: What is a good linear finite element? Interpolation, conditioning, anisotropy, and quality measures (preprint). University of California at Berkeley, vol. 73, p. 137 (2002)
33. Shewchuk, J.R.: Constrained delaunay tetrahedralizations and provably good boundary recovery. In: Eleventh International Meshing Roundtable (IMR), pp. 193–204 (2002)
34. Stainvas, I., Buda, Y.: Performance evaluation for curb detection problem. In: 2014 IEEE Intelligent Vehicles Symposium Proceedings, pp. 25–30. IEEE (2014)
35. Sui, L., Zhu, J., Zhong, M., Wang, X., Kang, J.: Extraction of road boundary from MLS data using laser scanner ground trajectory. Open Geosci. **13**(1), 690–704 (2021)
36. Sullivan, C.B., Kaszynski, A.: PyVista: 3D plotting and mesh analysis through a streamlined interface for the visualization toolkit (VTK). J. Open Source Softw. **4**(37), 1450 (2019). https://doi.org/10.21105/joss.01450
37. Szegedy, C., et al.: Going deeper with convolutions. In: Proceedings of the IEEE Conference on Computer Vision and Pattern Recognition, pp. 1–9 (2015)
38. Szegedy, C., Vanhoucke, V., Ioffe, S., Shlens, J., Wojna, Z.: Rethinking the Inception architecture for computer vision. In: Proceedings of the IEEE Conference on Computer Vision and Pattern Recognition, pp. 2818–2826 (2016)
39. Tagliasacchi, A., Delame, T., Spagnuolo, M., Amenta, N., Telea, A.: 3D skeletons: a state-of-the-art report. In: Computer Graphics Forum, vol. 35, pp. 573–597. Wiley Online Library (2016)
40. Weinmann, M., Jutzi, B., Mallet, C.: Semantic 3D scene interpretation: a framework combining optimal neighborhood size selection with relevant features. ISPRS Ann. Photogram. Remote Sens. Spatial Inf. Scie. **2**(3), 181 (2014). https://doi.org/10.5194/isprsannals-II-3-181-2014
41. Zhang, J., Zhao, H., Li, J.: TRS: transformers for remote sensing scene classification. Remote Sens. **13**(20), 4143 (2021)
42. Zhao, L., Yan, L., Meng, X.: The extraction of street curbs from mobile laser scanning data in urban areas. Remote Sens. **13**(12), 2407 (2021)
43. Zhou, Q.Y., Park, J., Koltun, V.: Open3D: a modern library for 3D data processing. arXiv:1801.09847 (2018)
44. Zhou, Z., Zheng, Y., Ye, H., Pu, J., Sun, G.: Satellite image scene classification via ConvNet with context aggregation. In: Hong, R., Cheng, W.-H., Yamasaki, T., Wang, M., Ngo, C.-W. (eds.) PCM 2018. LNCS, vol. 11165, pp. 329–339. Springer, Cham (2018). https://doi.org/10.1007/978-3-030-00767-6_31

Author Index

Printed in the United States
by Baker & Taylor Publisher Services